Science Library

Probability

WILEY SERIES IN PROBABILITY AND STATISTICS

Established by WALTER A. SHEWHART and SAMUEL S. WILKS

Editors: *Vic Barnett, Ralph A. Bradley, Nicholas I. Fisher, J. Stuart Hunter, J. B. Kadane, David G. Kendall, David W. Scott, Adrian F. M. Smith, Jozef L. Teugels, Geoffrey S. Watson*

A complete list of the titles in this series appears at the end of this volume

Probability

A Survey of the Mathematical Theory

Second Edition

JOHN W. LAMPERTI

Wiley Series in Probability and Statistics

A Wiley-Interscience Publication
JOHN WILEY & SONS, INC.
New York • Chichester • Brisbane • Toronto • Singapore

This text is printed on acid-free paper.

Copyright © 1996 by John Wiley & Sons, Inc.

All rights reserved. Published simultaneously in Canada.

Reproduction or translation of any part of this work beyond that permitted by Section 107 or 108 of the 1976 United States Copyright Act without the permission of the copyright owner is unlawful. Requests for permission or further information should be addressed to the Permissions Department, John Wiley & Sons, Inc., 605 Third Avenue, New York, NY 10158-0012.

Library of Congress Cataloging in Publication Data:
Lamperti, J. (John)
    Probability : a survey of the mathematical theory / John W. Lamperti. — 2nd ed.
      p. cm. — (Wiley series in probability and statistics)
    "A Wiley Interscience publication."
    Includes bibliographical references (p. –) and index.
    ISBN 0-471-15407-5 (alk. paper)
    1. Probabilities. I. Title. II. Series: Wiley series in probability and statistics. Probability and statistics.
QA273.L26 1996
519.2—dc20   96-9559

Printed in the United States of America

10 9 8 7 6 5 4 3 2 1

Contents

Preface   vii

1. Foundations   1

   1. Probability Spaces, 1
   2. Random Variables and Their Expectations, 5
   3. Independence, 12
   4. Conditional Expectations and Probabilities, 17
   5. When Do Random Variables Exist? 26

2. Laws of Large Numbers and Random Series   31

   6. The Weak Law of Large Numbers, 31
   7. The Weierstrass Approximation Theorem, 38
   8. The Strong Law of Large Numbers, 41
   9. The Strong Law—Continued, 44
  10. Convergence of Random Series, 52
  11. More on Independence; the 0–1 Law, 57
  12. The Law of the Iterated Logarithm, 61

3. Limiting Distributions and the Central Limit Problem   71

  13. Weak Convergence of Measures, 71
  14. The Maximum of a Random Sample, 78
  15. Characteristic Functions, 81
  16. The Central Limit Theorem, 94
  17. Stable Distributions, 100

18. Limit Distributions for Sums and Maxima, 110
  19. Infinitely Divisible Distributions, 118
  20. Recurrence, 125

4. The Brownian Motion Process   **131**

  21. Brownian Motion, 131
  22. The First Construction, 136
  23. Some Properties of Brownian Paths, 143
  24. Markov Processes, 150
  25. Brownian Motion and Limit Theorems, 157
  26. Brownian Motion and Classical Analysis, 165

Appendix: Essentials of Measure Theory   **175**

Bibliography   **181**

Index   **183**

List of Series Titles

Preface

The first edition of this book resulted from a one-semester graduate course I taught at Dartmouth College. The book, like the course, offered an overview of "classical" probability theory (up to around 1950), with results about sums of independent random variables occupying center stage. The book made some friends in its day, and I still hear occasionally from readers who found it helpful, or from teachers who wish to use it in a course.

That first edition has been out of print for several years, and I hope this new version will prove useful to teachers and students of our subject. It is fundamentally the same book, but entirely rewritten and, I hope, substantially improved. It has not been watered down. I have attempted to improve the exposition and, in places, to help the reader with better motivation and a few more examples. As in the original edition, I have tried to be honest but brief in the necessary use of measure theory, keeping the focus on the probabilistic ideas. Most importantly, perhaps, the book is still informal and short. A reader who finishes it will not know all there is to know about any part of the field; but, in compensation, that reader can begin section one with the expectation of finishing the book in a reasonable period of time, and then go on to current problems of his or her choosing.

Differences from the first edition include new sections on conditional probability in Chapter 1 and on the Markov concept in Chapter 4, plus a brief appendix summarizing necessary ideas and tools from measure theory. In Chapter 3 the reader will now find an account of the elegant parallel between limiting distributions for sums and for maxima of independent random variables. On the other hand, I have removed from Chapter 4 the sections about Markov transition functions which can be found in my book (or another) on stochastic processes and which belong more naturally there. (See the bibliography.) There are also many smaller differences in proofs, exposition, and problems.

As before, this book is not intended to provide anyone's first acquaintance with probability. (William Feller's classic *Introduction* is still my favorite for getting started in the field, but today there are also other good choices.) The other prerequisite, in addition to elementary probability, is knowledge of analysis at roughly beginning graduate level; Halsey Royden's *Real Analysis* contains all that's needed in this area and more, as do many other books. Of course, some willingness to think things through for oneself is also essential for best results!

The need to think for oneself pertains not only to the study of mathematics proper, but also to the ways in which mathematics and science are used. I believe that each of us has a measure of responsibility for the results of our work, a responsibility which cannot be simply passed along to employers or to governments. The need to consider the consequences is most evident for applied science and engineering—and, of course, probability theory has applications, for good or ill, in a great many areas of human activity. But the scientist's responsibility applies to the practice of "pure" mathematics as well. Through both teaching and research we are part of a collective enterprise with enormous social consequences, and the alleged "uselessness" of a particular theorem or concept does not shelter us from a share of personal involvement in how mathematics and science are put to work.

While working on this book I learned that the 1995 Nobel Prize for Peace had been awarded to the physicist Joseph Rotblat and the Pugwash Conferences on Science and World Affairs which he helped to found. In accepting the prize, Rotblat commented:

> I want to speak as a scientist, but also as a human being. From my earliest days I had a passion for science. But science, the exercise of the supreme power of the human intellect, was always linked in my mind with benefit to people. I saw science as being in harmony with humanity. I did not imagine that the second half of my life would be spent on efforts to avert a mortal danger to humanity created by science. [The danger is nuclear war.]
>
> At a time when science plays such a powerful role in the life of society, when the destiny of the whole of mankind may hinge on the results of scientific research, it is incumbent on all scientists to be fully conscious of that role, and conduct themselves accordingly. I appeal to my fellow scientists to remember their

responsibility to humanity. [Quoted from *The Bulletin of the Atomic Scientists* 52 (March/April 1996), pp. 26–28.]

I urge the readers of this book to remember Joseph Rotblat's appeal, and I hope we can work together to put it into practice.

JOHN W. LAMPERTI
Hanover, N.H.
April, 1996

A WORD ON NOTATION

Equations, theorems, and problems are numbered within each section simply as 1, 2, 3, etc. A reference to "Theorem n," for example, always means a theorem in the current section of the book, whereas "Theorem 15.3" refers to Theorem 3 in Section 15. References to problems and equations are treated the same way, so that "(8.4)" indicates the fourth numbered equation of Section 8.

Probability

CHAPTER ONE

Foundations

1. PROBABILITY SPACES

Let Ω be any nonempty set and suppose that $\mathcal{B}$ is a *Borel field* or, equivalently, a *σ-field* (pronounced "sigma field") of subsets of Ω. This means that $\mathcal{B}$ is a collection of subsets that contains the empty set $\varnothing$ and is closed under the formation of complements and of finite or countable unions of its members. Let P be a nonnegative function defined on $\mathcal{B}$ such that $P(\Omega) = 1$ and which is *countably additive* in the sense that

$$P\left(\bigcup_{n=1}^{\infty} A_n\right) = \sum_{n=1}^{\infty} P(A_n) \tag{1}$$

provided $A_n \in \mathcal{B}$ and $A_n \cap A_m = \varnothing$ for each $n \neq m$.

Definition 1. *Under these conditions P is a* probability measure *and the triple $(\Omega, \mathcal{B}, P)$ is a* probability space. *Sets belonging to $\mathcal{B}$ are called* events.

This definition generalizes the discrete probability spaces that suffice for many applications and are commonly used to introduce the subject.

Problem 1. *Suppose $\Omega = \{\omega_n\}$ is a finite or countably infinite set, and let $\mathcal{B}$ denote the collection of all its subsets. Assume that a nonnegative*

1

number p_n is assigned to each point $\omega_n \in \Omega$, with $\sum p_n = 1$. Define

$$P(A) = \sum_{\omega_n \in A} p_n \qquad (2)$$

for any set $A \subset \Omega$. Show that this defines a probability measure on $\mathcal{B}$, and that conversely all such measures arise in this way.

Beside the discrete spaces of Problem 1, examples include the unit interval $[0, 1]$ as Ω with the Borel sets and Lebesgue measure in the roles of $\mathcal{B}$ and P respectively, as well as the counterparts of these in higher dimensions; this example (in one dimension) is used as the model for "choosing a number at random" between 0 and 1. A bit more generally, any set S in R^n with positive, finite measure can be used as Ω with P defined by $P(A) = m(A)/m(S)$ for any measurable subset A of S, where m denotes Lebesgue measure in R^n. We can then speak of "choosing a point at random" from S, and this measure is called the *uniform distribution* on S.

Going a step further, we say that a nonnegative measurable function $f(x)$ on R^n is a *probability density* if its Lebesgue integral over the entire space R^n is equal to 1. The probability of a (measurable) set A is then defined to be the integral of f over A. This construction produces all the probability measures on R^n that are absolutely continuous with respect to Lebesgue measure. Elementary books often refer to this sort of probability space as the "continuous case"; its full definition supposes that the concepts of measurable sets and functions, and the construction of the Lebesgue measure in R^n, are already known. In many examples, of course, the density f is piecewise continuous or better, and the probability of "nice" sets can be calculated by ordinary Riemann integrals.

To derive more complicated probability measures, the following *extension theorem* is often used. (This result generalizes one approach to the construction of Lebesgue measure.) Suppose that $\mathcal{F}$ is a *field* or *Boolean algebra* of subsets of Ω; that is, $\mathcal{F}$ is a collection of subsets containing $\emptyset$ and closed under complementation and finite unions. Let P be a nonnegative function defined on $\mathcal{F}$ that has $P(\Omega) = 1$ and is *finitely additive;* that is, satisfies Eq. (1) for any finite collection of disjoint sets. Suppose that P also satisfies the following *continuity condition:*

If $A_n \in \mathcal{F}$, $A_{n+1} \subset A_n$, and $\bigcap_{n=1}^{\infty} A_n = \emptyset$, then $\lim_{n \to \infty} P(A_n) = 0$.

$$(3)$$

Let $\mathcal{B} = \mathcal{B}(\mathcal{F})$ be the smallest σ-field containing all the members of $\mathcal{F}$; this field is said to be *generated* by $\mathcal{F}$.[1]

Theorem 1. *Under the above conditions, there exists a unique probability measure on $(\Omega, \mathcal{B})$ that is an extension of P.*

We will not prove this theorem, which can be found in many books on measure and integration theory. [For instance, it appears in Royden (1988), Chapter 12, Section 2. The condition (3) appears there in a slightly different, but equivalent, form; see also Problem 2 below.] In some older works on probability such as Kolmogorov's *Foundations,* the term "probability space" means a triple $(\Omega, \mathcal{F}, P)$ satisfying the hypotheses of Theorem 1; since every such space extends uniquely to a probability space as defined here, the difference between that usage and ours is not important.

Problem 2. *Provided the function P is finitely additive, show that the continuity condition (3) is equivalent to the assumption that (1) (countable additivity) holds whenever the union of countably many disjoint sets from $\mathcal{F}$ happens to belong to $\mathcal{F}$. This shows that (3) is a necessary condition for the possibility of extending P to a measure on $\mathcal{B}(\mathcal{F})$.*

The case where Ω is the real line and $\mathcal{B}$ is the σ-field of Borel sets is particularly important.[2] Given a probability measure P on this space, the point function F defined by

$$F(t) = P(\{u : -\infty < u \leq t\}) \tag{4}$$

is called the *distribution function* of P and is easily seen to have these three properties:

1. F is nondecreasing;
2. F is continuous from the right;
3. $\lim_{x \to -\infty} F(t) = 0$ and $\lim_{x \to +\infty} F(t) = 1$.

---

[1] The intersection of any collection of σ-fields is again a σ-field; hence the intersection of all the σ-fields that contain $\mathcal{F}$ must be this smallest one. This simple argument parallels the proof that there exists a smallest subgroup or a smallest subspace of a vector space that contains a given set of elements.
[2] Reminder: the "Borel sets" of the line (or of R^n, or of any metric space) are those sets belonging to the σ-field generated by the open sets.

Conversely, if we are given any function F defined on the real line and satisfying properties 1, 2 and 3, then there exists a unique measure P on the Borel sets of the line that is related to F by (4). (Consequently, any function F satisfying the three conditions is called a distribution function.) To construct P, we first define $\mathcal{F}$ to be the field consisting of all finite unions of half-open intervals $(a, b]$, where $-\infty \le a < b \le \infty$. For each such interval $I = (a, b]$ we define $P(I) = F(b) - F(a)$, and for a disjoint union J of such intervals $P(J)$ equals the sum of the values of P for the intervals in J. This defines P as a function on $\mathcal{F}$; it is not hard to see that the value of $P(J)$ is not changed if the set $J \in \mathcal{F}$ is represented as a disjoint union of intervals in different ways. The plan, of course, is to extend the set function P as presently constituted from $\mathcal{F}$ to a countably-additive measure on $\mathcal{B}$ by means of Theorem 1; the only serious obstacle is verifying the continuity condition (3).

Problem 3. *Prove that (3) holds in the above situation. [Hint: Let A_n be a decreasing sequence of sets in $\mathcal{F}$ for which $P(A_n)$ does not tend to 0. Show that it is then possible to find compact sets $A'_n \subset A_n$ such that the sequence $\{A'_n\}$ has the finite-intersection property. It follows that the intersection of the A'_n, and hence of the A_n, cannot be empty.]*

There is a simple way to get the measure P from the distribution function if we assume the existence of Lebesgue measure m on $[0, 1]$. Let F be a distribution function that is continuous and strictly increasing, and for any Borel set A define $P(A) = m(F(A))$.

Problem 4. *Show that P is a probability measure and that (4) holds. How should these statements be modified if F has jumps and/or level intervals?*

Some authors require more of fields and measures before they award the title "probability space." One common additional postulate is that the space be *complete*: If $A \in \mathcal{B}$, $P(A) = 0$ and $A' \subset A$, then $A' \in \mathcal{B}$ also [and automatically $P(A') = 0$]. This assumption that subsets of events with probability 0 also are events with probability 0 agrees with the intuitive sense of "probability," and it is harmless in the sense that any Borel field can be "completed" with respect to a given measure P in a natural way. Moreover, some of the general methods for constructing measures lead automatically to complete spaces (see Royden).

There is, however, a disadvantage to including completeness as a postulate. At times we may need to consider more than one probability measure on the same σ-field, and while each one of them can be completed, it may not be possible to find a single enlarged σ-field on which all the measures are simultaneously defined and complete. For example, suppose P_1 is concentrated on a finite number of points of $[0, 1]$; then its completion is defined for *all* subsets of the interval. (Why?) If, by contrast, P_2 is Lebesgue measure on the Borel sets of $[0, 1]$, then its completion will be defined (only) for the field of Lebesgue-measurable sets and cannot be extended to all sets. In any event, although most of the measures we will encounter are going to be complete, we do not assume it as part of the definition of a probability space.

Problem 5. *Discuss the process of* completion. *That is, if $(\Omega, \mathcal{B}, P)$ is any probability space, show that there is a complete space $(\Omega, \mathcal{B}', P')$ such that $\mathcal{B}' \supset \mathcal{B}$ and P' is an extension of P. If $\mathcal{B}'$ is the smallest σ-field containing $\mathcal{B}$ as well as all subsets of sets of measure 0, show that P' is uniquely determined on $\mathcal{B}'$.*

2. RANDOM VARIABLES AND THEIR EXPECTATIONS

Let $(\Omega, \mathcal{B}, P)$ be a probability space and $X(\omega)$ a real-valued function on Ω. If X is Borel-measurable with respect to the σ-field $\mathcal{B}$—that is, if the set $X^{-1}(C) = \{\omega : X(\omega) \in C\}$ is a member of $\mathcal{B}$ for every Borel set C of real numbers (equivalently, for every open set or even for every open interval)—then X is a *random variable*. It is common for problems in pure or applied probability to be stated entirely in terms of the properties of one or more random variables, and sometimes the underlying probability space Ω is not mentioned at all.

Given any probability space, it is possible to construct a generalized Lebesgue integral over the space having the natural property that the integral of the indicator function of every set $A \in \mathcal{B}$ is just $P(A)$.[3] The concept and basic properties of such integrals are assumed familiar to the reader. (They are also reviewed briefly in the appendix.) We will use the notation

$$E(X) = \int_\Omega X(\omega)\, dP \tag{1}$$

---

[3] The *indicator function* $\mathbf{1}_A$ is defined by $\mathbf{1}_A(\omega) = 1$ for $\omega \in A$ and $\mathbf{1}_A(\omega) = 0$ otherwise.

for this integral; in probability theory it is called the *expected value* or *mean value* of X. [When we speak of $E(X)$ it is understood that the integral of $|X|$ is finite; otherwise the expected value does not exist.] The expected value, naturally, shares the basic properties of integrals from the general theory. First, it acts as a *linear functional* on the space of integrable random variables: If $X_1, \ldots, X_n$ are any random variables on $(\Omega, \mathcal{B}, P)$ with finite expectations and $c_1, \ldots, c_n$ are constants, then

$$E\left(\sum_{i=1}^{n} c_i X_i\right) = \sum_{i=1}^{n} c_i E(X_i). \tag{2}$$

The functional E is *nonnegative*, in the sense that $X \geq 0$ implies $E(X) \geq 0$. It obeys the *Schwarz inequality*:

$$|E(XY)| \leq [E(X^2) E(Y^2)]^{1/2}, \tag{3}$$

with equality if and only if $X = 0$ or $Y = cX$, either condition holding except possibly on a set of measure 0. The standard arguments that prove this inequality in finite-dimensional vector spaces also serve to prove it in the present context. Finally, the expectation functional is subject to the monotone and dominated convergence theorems (see the appendix).

In the case of a discrete probability space, all functions are measurable and hence qualify as random variables. In this situation the integral (1) over the space Ω reduces to a sum so that

$$E(X) = \sum_{n} X(\omega_n) p_n, \tag{4}$$

where $p_n = P(\{\omega_n\})$. There is, however, another way to compute $E(X)$:

$$E(X) = \sum_{m} x_m P(X = x_m), \tag{5}$$

where the sum now ranges over all real numbers x_m that are values taken on by the function X. It is easy to see that (4) and (5) are equivalent since the second sum is the same as the first one except that (possibly) blocks of terms in (4) collapse into single terms of (5). The availability of these alternative expressions for $E(X)$ is frequently useful.

There is an analogous alternative way to compute $E(X)$ in the general case. To derive it, we observe that a random variable X induces a probability measure P_X on the real line. The definition of this new measure for any

RANDOM VARIABLES AND THEIR EXPECTATIONS

Borel set C of reals is

$$P_X(C) = P(X^{-1}(C)). \tag{6}$$

The measure P_X is called the *distribution* of X; the corresponding point function F_X defined by

$$F_X(t) = P(\{\omega : X(\omega) \leq t\}) \tag{7}$$

is the *distribution function* of X and is clearly identical to the distribution function of the measure P_X as defined in Section 1. Using these concepts we have the desired generalization of (5): Provided that the integrals exist,

$$E(X) = \int_{R^1} t\, dP_X(t) = \int_{-\infty}^{\infty} t\, dF_X(t), \tag{8}$$

where the middle expression means the Lebesgue–Stieltjes integral of the identity function with respect to the measure P_X. Since $f(t) = t$ is continuous, this can be reinterpreted as a Riemann–Stieltjes integral, as in the final expression.

Why is this true? The first part of (8) is a special case of the following situation and of Theorem 1 below: Let $(\Omega', \mathcal{B}')$ be a *measurable space*; that is, a nonempty set together with a σ-field of subsets. Suppose that Φ is a mapping from Ω into Ω' which is measurable in the sense that $\Phi^{-1}(A') \in \mathcal{B}$ whenever $A' \in \mathcal{B}'$. Using this mapping plus the original probability measure P, a new measure P_Φ can be defined on $(\Omega', \mathcal{B}')$ just as the distribution of a random variable was defined above:

$$P_\Phi(A') = P(\Phi^{-1}(A')) \tag{9}$$

for any set $A' \in \mathcal{B}'$. Now suppose that X' is a (Borel) measurable function from Ω' to the real line, and let $X(\omega) = X'(\Phi(\omega))$ be the composition of the mappings Φ and X'.

Theorem 1. *The composite function X is a random variable on $(\Omega, \mathcal{B}, P)$ and*

$$E(X) = \int_{\Omega'} X'\, dP_\Phi, \tag{10}$$

where the existence of either side implies that of the other side.

Proof. The measurability of X is obvious. To prove (10), suppose first that X' is the indicator function of a set $A' \in \mathcal{B}'$; then the right side of (10) is just $P_\Phi(A')$. But in this case X is also an indicator function, that of $\Phi^{-1}(A')$, so $E(X) = P(\Phi^{-1}(A'))$. In view of the definition (9) these are equal, and (10) therefore holds for indicator functions. The extension to finite linear combinations—simple functions[4]—follows easily.

Now suppose only that $X' \geq 0$. Then there exists an increasing sequence of simple functions X'_n that tend to X', and by the monotone convergence theorem we have

$$\lim_{n \to \infty} \int_{\Omega'} X'_n \, dP_\Phi = \int_{\Omega'} X' \, dP_\Phi. \tag{11}$$

But the composite functions $X_n(\omega) = X'_n(\Phi(\omega))$ (again simple functions) are also increasing and tend to X, so again by monotone convergence we have

$$\lim_{n \to \infty} E(X_n) = E(X). \tag{12}$$

Since (10) holds for the functions X_n and X'_n, it follows from (11) and (12) that (10) also holds for X and X'.

Finally, the argument above can be applied separately to the positive and negative parts of X' to establish (10) in the general case. It is also clear that $E(X)$ exists if and only if X' is integrable with respect to P_Φ. □

To obtain Eq. (8)—the alternative calculation of the expected value—from this theorem, we take the real line R^1 as the space Ω' with the random variable X in the role of Φ; X' is now the identity function from R^1 into itself. More generally, we have this:

Corollary 1. *If X is any random variable and g is a Borel function on R^1, then*

$$E[g(X)] = \int_{R^1} g(u) \, dP_X(u) \tag{13}$$

provided either side exists.

Problem 1. *Prove this corollary.*

---

[4] An equivalent characterization of "simple functions" is that they take on only a finite number of values.

Remark. In many cases of interest the distribution P_X is absolutely continuous and has a density function $f(t) = F'(t)$ such that

$$P(a \le X \le b) = P_X([a,b]) = \int_a^b f(u)\,du; \tag{14}$$

in this case, the expected value, if it exists, can be calculated as

$$E(X) = \int_{-\infty}^\infty u\,f(u)\,du. \tag{15}$$

Elementary textbooks on probability and statistics often say that a random variable X is *continuous* if its distribution has a density satisfying (14), and then they use (15) to *define* the expected value of X. This works well enough until it becomes necessary to calculate quantities like $E(X^2)$ or, more generally, $E[g(X)]$. Using (15) as the definition would then require us to first find the density function of $g(X)$ in order to calculate its expectation. No one actually does that. Instead, these texts call [as in (13) above] for the calculation

$$E[g(X)] = \int_{-\infty}^\infty g(u)\,f(u)\,du. \tag{16}$$

The justification of (16), which requires Theorem 1 or something like it, is then sometimes ignored. (References to specific texts will be omitted to protect the guilty.)

In spite of the convenience of (13) or (16) for calculating expectations, it *is* sometimes necessary to find the distribution and the density of $g(X)$. This can be done using the distribution function F. For example, suppose that $Y = aX + b$, where $a > 0$ and b are constants. Then

$$P(Y \le t) = P\left(X \le \frac{t-b}{a}\right) = F\left(\frac{t-b}{a}\right), \tag{17}$$

and so if X has the density $f(t)$, the density of Y must be $a^{-1}f[(t-b)/a]$. When two random variables X and Y are related in this way by a linear change of scale, their distributions are said to be *of the same type*.

Problem 2. *Assuming that a random variable X has the density f, find the density functions of X^2 and X^3.*

Example. The most important distribution of probability theory is the so-called *normal* or *Gaussian*, which in its standard form has the distribution function

$$N(x) = \frac{1}{\sqrt{2\pi}} \int_{-\infty}^{x} e^{-t^2/2}\, dt. \tag{18}$$

There is a well-known trick, elementary but not obvious, which is used to show that $N(\infty) = 1$. (This is given in most calculus texts.) Noting that the density function in (18) is even, it is clear from (15) that the expectation of a standard normal random variable must be 0; by (16) we see that the same holds for any odd power of Z.[5] Using repeated integration by parts, we find that

$$E(Z^2) = 1, \quad E(Z^4) = 3, \quad \ldots, \quad E(Z^{2k}) = 1 \cdot 3 \cdot 5 \cdots k. \tag{19}$$

A random variable related to Z by a linear transformation is also called "normal." Suppose that $Y = aZ + b$, where $a > 0$; then by (17) above Y has the density

$$\frac{d}{dt} P(Y \le t) = \frac{1}{\sqrt{2\pi}\, a}\, e^{-(t-b)^2/2a^2}. \tag{20}$$

Without calculations it is easy to see that $E(Y) = b$ and that $E(Y^2) = a^2 + b^2$.

Problem 3. *Verify* (19) *and find* $E(Y^k)$ *for any positive integer* k.

Many of the usual notions of convergence of a sequence of functions play a role in probability theory, but it is customary to disguise them with different names. If X_n is a sequence of random variables defined on a probability space $(\Omega, \mathcal{B}, P)$ such that the set $\{\omega : \lim X_n(\omega) \text{ exists}\}$ has P-measure equal to 1, the sequence is said to *converge almost surely* (abbreviated a.s.); this, of course, is the same as "convergence almost everywhere" (i.e., except on a set of measure zero) from real variable theory.

Problem 4. *Show that if* X_n *converges almost surely, the function* $X(\omega)$ *that equals* $\lim X_n(\omega)$ *when the limit exists and* 0 *otherwise is a random variable.*

---

[5]It is a common convention to use the symbol Z for any random variable with the distribution function (18).

The concept of convergence in measure becomes in our setting

$$\lim_{n\to\infty} P(\{\omega : |X_n(\omega) - X(\omega)| > \epsilon\}) = 0 \qquad (21)$$

for each $\epsilon > 0$, and this is now called *convergence in probability*. Finally, if

$$\lim_{n\to\infty} E(|X_n - X|^\alpha) = 0 \qquad (22)$$

for some $\alpha > 0$, we have *convergence in the mean* of order α. It is not difficult to show that either convergence in the mean (of any positive order) or convergence a.s. implies convergence in probability, and that there is no other implication among the different modes. All three of them will be used in Chapter 2 while discussing "laws of large numbers."

Problem 5. *Verify the next to last sentence.*

We conclude this section with a brief mention of another axiom that is sometimes included in the definition of a probability space—the requirement that the space be *perfect*. We have defined in (6) above the measure P_X induced on the Borel sets of the real line by a random variable X. The measure P_X can always be completed as in Problem 1.5, and in the process its definition is extended to a σ-field that may be much larger than the field of Borel sets. Even on this larger field the values of P_X can be reconstructed from the distribution function $F_X(t) = P(X \le t)$. However, there is no reason why (6) might not assign a value to P_X for even more sets of real numbers if the field $\mathcal{B}$ is rich enough; a perfect measure is one for which this possibility does not occur for any random variable X. This axiom is discussed and used in Gnedenko and Kolmogorov (1968), for example, and the appendix by J. L. Doob (added to that book in its English edition) is worth reading for a discussion of several foundation questions as well as for notes on perfect measures.

As with completeness, most measures we are going to encounter will be perfect. But we do not assume this condition, nor will we stop to verify it even when it holds in some particular case. Instead, we usually adopt the attitude that if X is a random variable, then $P(X \in C)$ is of interest only for those sets C whose P_X measure is determined by the distribution function of X, and the possible $\mathcal{B}$–measurability of $\{\omega : X(\omega) \in C\}$ for other sets C will be ignored.

3. INDEPENDENCE

Let $(\Omega, \mathcal{B}, P)$ be any probability space, and let $A_1, \ldots, A_n$ be *events*; that is, sets belonging to $\mathcal{B}$.

Definition 1. *The* events $A_1, \ldots, A_n$ *are called* independent *provided every subcollection of them satisfies*

$$P(A_{i_1} \cap A_{i_2} \cap \cdots \cap A_{i_k}) = P(A_{i_1})P(A_{i_2}) \cdots P(A_{i_k}). \tag{1}$$

Random variables $X_1, \ldots, X_n$ *are* independent *if the events*

$$A_k = \{\omega : X_k(\omega) \in S_k\} \tag{2}$$

satisfy (1) *for every choice of Borel sets* $S_1, \ldots, S_n$ *in* R^1. *Infinite collections of random variables are said to be independent when all finite subcollections satisfy the above condition.*

As Kolmogorov and others have remarked, it is the concept of independence more than anything else that gives probability theory a life and flavor of its own and separates it from other branches of analysis.

There is an important relation between independence of random variables and their expectations. The following result will be used repeatedly.

Theorem 1. *Let X and Y be independent random variables. Then $E(XY)$ exists if both $E(X)$ and $E(Y)$ exist, and in that case we have*

$$E(XY) = E(X)E(Y). \tag{3}$$

Conversely, if $E(XY)$ exists and neither X nor Y vanishes a.s., then $E(X)$ and $E(Y)$ exist as well.

Proof. Suppose that both X and Y are simple functions, that is, functions that take only finitely many values. Then there are two families of disjoint sets $\{A_i\}$ and $\{B_j\}$ and corresponding constants a_i and b_j such that

$$X(\omega) = \sum_{i=1}^{n} a_i \mathbf{1}_{A_i}(\omega) \quad \text{and} \quad Y(\omega) = \sum_{j=1}^{m} b_j \mathbf{1}_{B_j}(\omega). \tag{4}$$

Since X and Y are independent, we can infer that

$$P(A_i \cap B_j) = P(A_i)P(B_j)$$

INDEPENDENCE

for each i and j, and from this (3) follows easily:

$$E(XY) = E\left(\sum_i a_i \mathbf{1}_{A_i} \sum_j b_j \mathbf{1}_{B_j}\right)$$

$$= E\left(\sum_{i,j} a_i b_j \mathbf{1}_{A_i \cap B_j}\right) = \sum_{i,j} a_i b_j P(A_i \cap B_j)$$

$$= \left(\sum_i a_i P(A_i)\right)\left(\sum_j b_j P(B_j)\right) = E(X)E(Y). \quad (5)$$

Of course all the expectations necessarily exist for simple functions.

Now let X and Y be nonnegative and independent, but not necessarily simple. Define

$$X_n(\omega) = \begin{cases} \dfrac{i}{2^n} & \text{if } \dfrac{i}{2^n} \leq X(\omega) < \dfrac{i+1}{2^n}; \\ 0 & \text{otherwise} \end{cases} \quad (6)$$

for $i = 0, 1, \ldots, n2^n$, and construct Y_n similarly from Y. Then X_n and Y_n are simple and our construction ensures that they are independent;[6] it follows that (5) holds for these functions. Definition (6) implies as well that the $\{X_n\}$ increase monotonically to the limit X, and so by the monotone convergence theorem

$$\lim_{n \to \infty} E(X_n) = E(X). \quad (7)$$

The same things are of course true for $\{Y_n\}$ and Y. Finally, the products $X_n Y_n$ are also increasing to XY, so that $E(X_n Y_n)$ tends to $E(XY)$. Putting all this together, we see that (3) must hold for X and Y. Existence if and only if is also a consequence of the monotone convergence theorem, since if $E(XY) < \infty$ then both $E(X_n)$ and $E(Y_n)$ must be bounded—unless one or the other is always equal to 0.

Finally, we must remove the restriction that X and Y are nonnegative. Let $X^+ = \max(X, 0)$ and $X^- = \max(-X, 0)$, so that $X = X^+ - X^-$. It is easy to see that X is integrable [i.e., $E(X)$ exists] if and only if both X^+ and X^- are integrable. Perform the same decomposition on Y. From the

---

[6] It doesn't work to choose arbitrary sequences of simple functions increasing to X and to Y because, in general, X_n and Y_n will not be independent.

independence of X and Y it follows that the random variables X^+ and Y^+ are independent, as are the other three possible combinations; the proof is left as an exercise. Therefore we have

$$E(X^{\pm}Y^{\pm}) = E(X^{\pm})E(Y^{\pm}). \tag{8}$$

Now assume that $E(X)$ and $E(Y)$ both exist; then using (8) we have

$$\begin{aligned} E(XY) &= E[(X^+ - X^-)(Y^+ - Y^-)] \\ &= E(X^+Y^+) - E(X^+Y^-) - E(X^-Y^+) + E(X^-Y^-) \\ &= E(X^+)E(Y^+) - E(X^+)E(Y^-) - E(X^-)E(Y^+) + E(X^-)E(Y^-) \\ &= [E(X^+) - E(X^-)][E(Y^+) - E(Y^-)] = E(X)E(Y). \end{aligned} \tag{9}$$

In this process we have exhibited XY as the sum of four random variables with finite expectations, so we know that $E(XY)$ exists.

The only remaining step is to prove the converse, so assume that $E(XY)$ exists; then both $(XY)^+$ and $(XY)^-$ are integrable. But $(XY)^+ = X^+Y^+ + X^-Y^-$ and $(XY)^- = X^+Y^- + X^-Y^+$ (all nonnegative), so the four quantities $E(X^{\pm}Y^{\pm})$ must be finite. Now suppose that $E(X)$ does not exist. Then either $E(X^+)$ or $E(X^-)$ must be infinite; assume the former. But both $E(X^+Y^+)$ and $E(X^+Y^-)$ are finite. Using the results from the case of nonnegative variables, we conclude that both Y^+ and Y^- must be zero a.s. A similar argument works if it should be $E(X^-)$ that is infinite, so this completes the proof. $\square$

Problem 1. *Show that if X and Y are independent, then so are X^+ and Y^+ (as well as the other three possible combinations).*

Naturally we would like to extend Theorem 1 from two random variables to any finite number. There is a deceptively easy argument: Suppose (for the case when $n = 3$) that X, Y, and Z are independent and integrable. Then by a repeated application of the theorem,

$$E(XYZ) = E[X(YZ)] = E(X)E(YZ) = E(X)E(Y)E(Z).$$

The difficulty here is that we have not shown that X and (YZ) are independent (although they are). We will clarify this point (and more) in Section 11, but in the meantime a different approach is needed.

Corollary 1. *The conclusions of the theorem hold for any (finite) collection of $n \geq 2$ independent random variables.*

Problem 2. *Prove the corollary by adapting the method used above for the product of two random variables.*

In Section 2 we described how one random variable can be studied in terms of its distribution, that is, the measure it induces on the real line. A similar approach is possible, and important, in the case of two or more. If X and Y are random variables, independent or not, they generate a mapping $Z(\omega) = (X(\omega), Y(\omega))$ from Ω into R^2. This mapping is measurable (see Problem 3 below) and so it induces a measure $P_{X,Y}$ on the Borel sets of R^2 which is defined by (2.9) with Z in the role of Φ. This measure is called the *joint distribution* of X and Y. There is also a *joint distribution function*

$$F_{X,Y}(s,t) = P(\{\omega : X \leq s \text{ and } Y \leq t\}) \tag{10}$$

from which the measure $P_{X,Y}$ can be reconstructed, and when $P_{X,Y}$ is absolutely continuous with respect to Lebesgue measure there is a *joint density* $f(s,t)$ such that

$$P_{X,Y}(C) = \int_C f(s,t)\,ds\,dt. \tag{11}$$

Just as in the case of one random variable, the joint distribution (with or without the existence of a density) can be used to calculate expected values. Suppose that $g(s,t)$ is a Borel function on R^2; then, of course, $g(X,Y)$ is a random variable. Theorem 2.1 can now be applied with g in the role of the second mapping X'; the result is

$$E[g(X,Y)] = \int_{R^2} g(s,t)\,dP_{X,Y}(s,t), \tag{12}$$

a formula used often. In particular, choosing $g(s,t) = st$ provides another way to compute the expected value of the product XY, since (12) then becomes

$$E(XY) = \int_{R^2} st\,dP_{X,Y}(s,t). \tag{13}$$

These remarks all generalize to the case of n random variables.

Problem 3. *Show that $Z^{-1}(C) \in \mathcal{B}$ for every Borel set C in R^2.*

If the random variables X and Y of the previous paragraph are independent, then it follows directly from the definition that their joint distribution

$P_{X,Y}$ is the *product measure* formed from the individual distributions P_X and P_Y. (The joint distribution function is then the product of the individual distribution functions, and the same is true for the joint density if one exists.) In this situation, Fubini's theorem and related tools can be applied. In particular, another (perhaps shorter) proof of Theorem 1 can be given.

Problem 4. *Use Fubini's theorem and (13) to prove Theorem 1.*

Problem 5. *Use similar methods based on Theorem 2.1 and joint distributions to extend the result of Theorem 1 to n random variables.*

Much of traditional probability theory, and much of this book, is concerned with sums of two or more independent random variables. It is clear that the joint distribution of the random variables, and hence the distribution of their sum, is determined by the distributions of the individual variables. Our next result shows how.

Theorem 2. *Let X and Y be independent random variables having the distribution functions F and G, respectively. Then the distribution function of their sum is given by*

$$P(X + Y \le t) = \int_{-\infty}^{\infty} F(t - v)\, dG(v) = \int_{-\infty}^{\infty} G(t - u)\, dF(u). \qquad (14)$$

Proof. For fixed t, define the function

$$g(u, v) = \begin{cases} 1 & \text{if } u + v \le t; \\ 0 & \text{otherwise,} \end{cases} \qquad (15)$$

and apply Eq. (12). The left-hand side is then just $P(X + Y \le t)$. By independence $P_{X,Y}$ is a product measure, and so Fubini's theorem can be used to express the right-hand side as an iterated integral in either of two ways:

$$\begin{aligned}
\int_{R^2} g\, dP_{X,Y} &= \int_{-\infty}^{\infty} \left(\int_{-\infty}^{\infty} g(u, v)\, dP_X(u) \right) dP_Y(v) \\
&= \int_{-\infty}^{\infty} \left(\int_{-\infty}^{\infty} g(u, v)\, dP_Y(v) \right) dP_X(u).
\end{aligned}$$

Using the definition of g the two iterated integrals easily reduce to the two Lebesgue–Stieltjes integrals in (14), and so the theorem is proved. $\square$

Corollary 2. *Suppose the distribution of X is absolutely continuous, with density $f(u)$. Then the distribution of $X + Y$ is also absolutely continuous, and its density is given a.e. by*

$$\frac{d}{dt} P(X + Y \leq t) = \int_{-\infty}^{\infty} f(t - v) \, dG(v). \tag{16}$$

Proof. By Fubini's theorem, we have

$$\int_{-\infty}^{t} \left(\int_{-\infty}^{\infty} f(u - v) \, dG(v) \right) du =$$

$$\int_{-\infty}^{\infty} \left(\int_{-\infty}^{t} f(u - v) \, du \right) dG(v) = \int_{-\infty}^{\infty} F(t - v) \, dG(v).$$

The result is $P(X + Y \leq t)$ by (14), and so we have exhibited this distribution function as the integral of the right-hand side of (16). □

Problem 6. *Suppose that Y and Y' are independent random variables each having a normal distribution with density given by (2.20), where the constants are (a, b) and (a', b'), respectively. Using Corollary 2, show that the sum $Y + Y'$ also has a normal distribution, and find the corresponding constants a'' and b''.*

Remark. The integral in (14), in either of its two equivalent forms, is called the *convolution* of the distribution functions F and G; we will indicate it with a star as in $F \star G$. Clearly, many statements about sums of independent random variables can be rephrased as statements about convolutions of distribution functions or densities on R^1. This suggests that Fourier analysis could be an important tool for studying such sums; that this idea is correct will be amply demonstrated in Chapter 3.

4. CONDITIONAL EXPECTATIONS AND PROBABILITIES

If A and B are any two events with $P(A) > 0$, the *conditional probability of B given A* is written $P(B \mid A)$ and defined as

$$P(B \mid A) = \frac{P(B \cap A)}{P(A)}. \tag{1}$$

Notice that $P(B \mid A) = P(B)$ if and only if A and B are independent. Considered as a function of B, the right-hand side of (1) defines a probability measure on $(\Omega, \mathcal{B}, P)$, one which puts probability 1 on the set A. The integral (if it exists) of any random variable Y with respect to this measure is called the *conditional expectation* of Y given A:

$$E(Y \mid A) = \int_\Omega Y(\omega) P(d\omega \mid A). \tag{2}$$

The integral in (2) could, of course, be taken over A rather than over Ω, and it is possible to think of the conditional expectation $E(Y \mid A)$ as the average value of Y over the set A. All the discussion in the preceding sections can be applied to conditional probability spaces; in particular, we define the *conditional independence given A* of events or of random variables by using the definition of Section 3 with the new (conditional) measure. The reader is assumed to be somewhat familiar with these ideas.

Now suppose that X is any discrete random variable, taking the values $t_1, t_2, \ldots$ with probabilities $p_1, p_2, \ldots$, respectively, where $p_k > 0$. Since each event $\{X = t_k\}$ has positive probability, we can define the conditional expectation

$$E(Y \mid X = t_k) = \int_\Omega Y(\omega) P(d\omega \mid X = t_k) \tag{3}$$

as above for any k. The left side of (3) depends on which value t_k of X we are considering, so we can think of $E(Y \mid X)$ as a *function of X* and hence as a random variable. This random variable has a constant value on each set $\{X = t_k\}$, namely the average of Y over that set, and thus we have

$$\int_{\{X=t_k\}} E(Y \mid X) \, dP = E(Y \mid X = t_k) P(X = t_k) = \int_{\{X=t_k\}} Y \, dP. \tag{4}$$

Property (4), plus the fact that $E(Y \mid X)$ is a function of X, will be the basis for defining conditional expectations in much greater generality.

In many applications, conditional probabilities are simpler than unconditional ones, and they can be used as building blocks to find other quantities of interest. A useful tool is the fact that, for any X,

$$E[E(Y \mid X)] = E(Y) \tag{5}$$

whenever $E(Y)$ exists. In the discrete case under discussion so far, (5) is easily verified by summing (4) over all values of k. We can think of the conditional expectation as a smoothing of Y that replaces the original values

CONDITIONAL EXPECTATIONS AND PROBABILITIES   19

by their averages over the sets where X takes a given value. Equation (5) states that averaging this smoothed version of Y over the whole space Ω gives the same result as if Y had been averaged all at once.

Example 1. Suppose that an urn contains a fixed number N of balls, and that a random number X of those balls are red; the rest are white. A ball is drawn at random, and we seek the probability that it is red. The conditional probability of drawing a red ball given that $X = k$ is clearly k/N; hence by (5)

$$P(\text{red}) = \sum_{k=0}^{N} P(\text{red} \mid X = k) P(X = k) = \sum_{k=0}^{N} \frac{k}{N} P(X = k) = \frac{E(X)}{N}.$$

[More precisely, we applied (5) to the indicator function of the event {draw a red ball}. Probabilities can always be rewritten as expectations in this way.] This technique, using (5), is sometimes called *conditioning*.

Problem 1. *From the urn of the above example, a ball is drawn at random, replaced, and then another ball is randomly drawn. Show that the events of obtaining a red ball on the first and second draws are independent if and only if X is a.s. constant. [Hint: First show that the probability of two red balls is $E(X^2)/N^2$. Independence thus requires that $E(X^2) = E(X)^2$.]*

Example 2 (Random Sums). Let $X_1, X_2, \ldots$ be independent random variables with the same distribution, and let M be a random variable taking the values $0, 1, 2, \ldots$ that is independent of the $\{X_k\}$. We wish to study the sum

$$S = \sum_{k=0}^{M} X_k. \tag{6}$$

In particular, we seek $E(S)$. By conditioning on the value of M, we have

$$E(S) = E\left[E(S \mid M)\right] = \sum_{k=0}^{\infty} E(S \mid M = k) P(M = k)$$

$$= \sum_{k=0}^{\infty} k E(X) P(M = k) = E(X) E(M) \tag{7}$$

provided both expectations exist. In deriving (7), we argue that because of the independence of $\{X_k\}$ and M, the distribution of S given $M = k$

is simply the distribution of the sum of k random variables having the common distribution of the original X's.

To generalize the concepts just described for discrete X, we will work with conditional expectations.[7] We have observed that $E(Y \mid X)$ is a random variable that is a function of X. We notice that the *values* of the variable X play no role; all that matters is the way in which X decomposes the probability space. In other words, what counts is the collection of sets that are inverse images under X of the Borel sets in R^1. This collection of sets is a sub σ-field of $\mathcal{B}$ called the σ-field *generated by X*; it will be denoted $\mathcal{B}(X)$.[8] Instead of stating that the conditional expectation is a function of X, we assume that it is a random variable which is measurable with respect to the σ-field $\mathcal{B}(X)$. In addition, something like the equality of the first and third expressions in (4) is needed to characterize the conditional expectation; we will assume that such an equation must hold with the set $\{X = t_k\}$ replaced by any set belonging to $\mathcal{B}(X)$. Let us summarize.

Definition 1. *A random variable Z is said to be a* version *of the conditional expectation of Y given X [denoted $E(Y \mid X)$] provided it satisfies these conditions:*

1. *Z is measurable with respect to the σ-field $\mathcal{B}(X)$, and*
2. *$\int_A Z \, dP = \int_A Y \, dP$ for any set $A \in \mathcal{B}(X)$.*

If Y is the indicator function of a measurable set C, then Z is called the conditional probability *of C given X.*

Before we examine the properties of the object just defined, it is convenient to generalize it even more. In numerous applications we need conditional expectations with respect to not one but several random variables, say $X_1, \ldots, X_n$. This merely requires a change in the σ-field $\mathcal{B}(X)$ used in Definition 1. In fact, any collection $\mathcal{X}$ of random variables on the probability space $(\Omega, \mathcal{B}, P)$ generates a Borel field $\mathcal{B}(\mathcal{X})$ of sets, which is defined as the smallest subfield of $\mathcal{B}$ with respect to which each $X \in \mathcal{X}$ is measurable. [$\mathcal{B}(\mathcal{X})$ exists for the same reason as did $\mathcal{B}(\mathcal{F})$ in Section 1.] To

---

[7] The general theory of conditional expectations has many applications, but in this book it will be needed only in Section 24.
[8] Alternatively, $\mathcal{B}(X)$ can be described as the smallest σ-field with respect to which X is measurable. This terminology extends the notation $\mathcal{B}(\mathcal{F})$ in Section 1, where $\mathcal{F}$ was a collection of sets.

define $E(Y \mid X_1, \ldots, X_n)$ or $E(Y \mid \mathcal{X})$, we simply replace $\mathcal{B}(X)$ in Definition 1 by the field $\mathcal{B}(X_1, \ldots, X_n)$ or $\mathcal{B}(\mathcal{X})$ generated by all the conditioning variables we wish to use.

By now, however, it is becoming clear that the natural object of study is a conditional expectation with respect to a general σ-field $\mathcal{B}'$, whether or not that field was derived from some collection of random variables. Hence our working definition will be the following.

Definition 2. *Let Y be any random variable with $E(|Y|) < \infty$, and suppose that $\mathcal{B}'$ is any sub σ-field of $\mathcal{B}$. A random variable Z is a* version of the *conditional expectation of Y given $\mathcal{B}'$ [denoted $E(Y \mid \mathcal{B}')$], provided Z satisfies the conditions:*

1. *Z is measurable with respect to the σ-field $\mathcal{B}'$, and*
2. *$\int_A Z \, dP = \int_A Y \, dP$ for any set $A \in \mathcal{B}'$.*

If Y is the indicator function of a measurable set C, then Z is a version of the conditional probability *of C given $\mathcal{B}'$.*

Several questions must now be settled. Does $E(Y \mid \mathcal{B}')$ always exist? Is it unique? What are its most important properties? And finally, how does it relate to more elementary (and more intuitive) concepts of conditional probability and expectation? We will look at these matters in turn.

Theorem 1. *For any Y with $E(|Y|) < \infty$, there exist random variables Z that satisfy the conditions of Definition 2 and so are versions of $E(Y \mid \mathcal{B}')$. Any two such random variables must be equal a.s.*

Proof. Assume $E(|Y|) < \infty$ and, for the moment, that $Y \geq 0$. Then the set function

$$m(A) = \int_A Y(\omega) \, dP$$

defines a measure on $\mathcal{B}$ that is absolutely continuous with respect to P. We consider this measure restricted to sets belonging to the σ-field $\mathcal{B}'$ and apply the Radon–Nikodym theorem. The theorem asserts that $m(A)$ can be represented by a density function, that is, that

$$m(A) = \int_A Z(\omega) \, dP \quad \text{for all} \quad A \in \mathcal{B}',$$

where Z is measurable with respect to $\mathcal{B}'$. The function Z thus satisfies conditions 1 and 2 of Definition 2 and so is a version of the conditional expectation $E(Y \mid \mathcal{B}')$. If $Y \geq 0$ does not hold, we can apply this argument to both Y^+ and Y^- and combine the results.

Now let us assume that Z_1 and Z_2 are both versions of the conditional expectation. The set $A = \{\omega : Z_1(\omega) > Z_2(\omega)\}$ belongs to $\mathcal{B}'$, and so by property 2 we have

$$\int_A Z_1 \, dP = \int_A Y \, dP = \int_A Z_2 \, dP,$$

which yields

$$\int_A (Z_1 - Z_2) \, dP = 0.$$

Since the integrand is strictly positive on A, this implies $P(A) = 0$. Interchanging Z_1 and Z_2 we also have $P(Z_1 < Z_2) = 0$, so $Z_1 = Z_2$ a.s. and the proof of Theorem 1 is complete. □

In the less general situation where it is assumed that $E(Y^2) < \infty$, a different existence proof can be given that adds geometric insight. We consider the Hilbert space $L_2(\Omega, \mathcal{B}, P)$ and note that those functions in L_2 which are $\mathcal{B}'$-measurable form a closed subspace M. (The reader for whom such arguments are new should think through why this is so.) Let $\mathbf{P}_M$ denote the operator of perpendicular projection into this subspace. Then we will show that

$$\mathbf{P}_M Y = E(Y \mid \mathcal{B}') \qquad \text{(a.s.)}. \tag{8}$$

The verification is simple. Since it is an element of the subspace M, $\mathbf{P}_M Y$ must by definition be $\mathcal{B}'$-measurable. Now suppose that $A \in \mathcal{B}'$, so that the indicator function $\mathbf{1}_A \in M$. Then

$$\int_A Y \, dP = \int_\Omega Y \mathbf{1}_A \, dP = (Y, \mathbf{1}_A) = (Y, \mathbf{P}_M \mathbf{1}_A)$$

$$= (\mathbf{P}_M Y, \mathbf{1}_A) = \int_A \mathbf{P}_M Y \, dP,$$

which verifies 2. Here we have used (only) the facts that a projection operator acts as the identity on its range and is self-adjoint.

The most common and useful properties of conditional expectation will now be stated as a theorem for future reference.

Theorem 2. *Suppose $E(|Y|) < \infty$ and $\mathcal{B}'$ is a subfield of $\mathcal{B}$. Then:*

1. *The conditional expectation $E(Y \mid \mathcal{B}')$ is linear in Y.*
2. *Conditional expectation is nonnegative: If $Y \geq 0$ a.s. then $E(Y \mid \mathcal{B}') \geq 0$ a.s. Consequently,*

$$|E(Y \mid \mathcal{B}')| \leq E(|Y| \mid \mathcal{B}') \quad \text{a.s.}$$

3. *If Z is $\mathcal{B}'$-measurable and $E(|YZ|) < \infty$, then Z can be taken outside the conditional expectation:*

$$E(YZ \mid \mathcal{B}') = Z\, E(Y \mid \mathcal{B}') \quad \text{a.s.}$$

In particular, if Y is itself $\mathcal{B}'$-measurable then $E(Y \mid \mathcal{B}') = Y$ a.s.

4. *Suppose that $\mathcal{B}''$ is a sub σ-field of $\mathcal{B}'$. Then*

$$E[E(Y \mid \mathcal{B}') \mid \mathcal{B}''] = E(Y \mid \mathcal{B}'') \quad \text{(a.s.)}$$

Proof. All four properties can be verified from the definition of conditional expectation and properties of integrals. For example, to check property 4—which generalizes the conditioning equation (5)—we note that the left side is $\mathcal{B}''$-measurable by definition and so we must show that

$$\int_A E[E(Y \mid \mathcal{B}') \mid \mathcal{B}'']\, dP = \int_A Y\, dP \tag{9}$$

for any set $A \in \mathcal{B}''$. But since $A \in \mathcal{B}''$, we have first that

$$\int_A E[E(Y \mid \mathcal{B}') \mid \mathcal{B}'']\, dP = \int_A E(Y \mid \mathcal{B}')\, dP,$$

and because in addition $A \in \mathcal{B}'$ (since $\mathcal{B}'' \subset \mathcal{B}'$), we also have

$$\int_A E(Y \mid \mathcal{B}')\, dP = \int_A Y\, dP.$$

Combining these facts yields (9).

In the case when $E(Y^2) < \infty$, a more intuitive proof of property 4 consists of noting that the left side represents the projection of Y first into the subspace of all $\mathcal{B}'$-measurable functions, followed by a projection of the result into the sub-subspace of all $\mathcal{B}''$-measurable functions. The right side is the projection of Y directly into the smaller subspace. It is geometrically evident, and easy to verify in any Hilbert space, that these two operations give the same end result.

We will leave the three other properties for the reader.

Problem 2. *Verify result 1 of Theorem 2.*

Problem 3. *Verify 2. [Hint: Consider $A = \{\omega : E(Y \mid \mathcal{B}') < 0\}$.]*

Problem 4. *Verify property 3. [Hint: Start with the case when Z is the indicator function of a set $B \in \mathcal{B}'$.]* □

Finally, we return to more solid ground. Let us assume that the random variables X and Y have a joint density function $f(s, t)$ that is continuous and positive on R^2. (Weaker conditions will suffice.) Then the elementary definition of conditional expectation given in many texts is

$$E(Y \mid X = s) = \frac{\int_{-\infty}^{\infty} t f(s, t) \, dt}{\int_{-\infty}^{\infty} f(s, t) \, dt} = g(s). \tag{10}$$

This formula (fortunately) agrees with our general definition.

Proposition 1. *The random variable $g(X(\omega))$ is a version of $E(Y \mid X)$.*

Proof. Obviously $g(X)$ is measurable with respect to $\mathcal{B}(X)$. We must then verify that for any set $A \in \mathcal{B}(X)$,

$$\int_A g(X) \, dP = \int_A Y \, dP. \tag{11}$$

To do this, both integrals can be moved from Ω to R^2 by means of the mapping $Z(\omega) = (X(\omega), Y(\omega))$. First notice that A must be of the form $\{\omega : X \in C\}$ for some Borel set C of reals. Then using (3.12) plus the existence of a joint density, we have for the right side of (11)

$$\int_A Y \, dP = \int_\Omega Y \mathbf{1}_A \, dP = \int_{R^2} t \, \mathbf{1}_C(s) f(s, t) \, ds \, dt. \tag{12}$$

Given the same treatment, the left side becomes

$$\int_A g(X) \, dP = \int_\Omega g(X) \mathbf{1}_A \, dP = \int_{R^2} g(s) \mathbf{1}_C(s) f(s, t) \, ds \, dt. \tag{13}$$

Substituting the definition of the function g from (10) and using the equality of double and iterated integrals, we see easily that the right-hand sides of (12) and (13) are equal; this verifies (11) and proves the proposition. □

Similarly, it can be shown that if the random variables Y and $X_1, \ldots, X_n$ have the positive and continuous joint density $f(s, t_1, \ldots, t_n)$ and we define

$$g(t_1, \ldots, t_n) = \frac{\int_{-\infty}^{\infty} s f(s, t_1, \ldots, t_n) \, ds}{\int_{-\infty}^{\infty} f(s, t_1, \ldots, t_n) \, ds} \qquad (14)$$

then $g(X_1, \ldots, X_n)$ is a version of $E(Y \mid X_1, \ldots, X_n)$.

Example 3 (Joint Normal Distributions).[9] The random variables $X_0, X_1, \ldots, X_n$ have a nondegenerate *multivariate normal* (or *Gaussian*) distribution with means 0, provided they have a joint density of the form

$$f(t_0, t_1, \ldots, t_n) = K \exp\left(-\frac{1}{2} \sum_{i,j=0}^{n} d_{ij} t_i t_j\right), \qquad (15)$$

where $[d_{ij}]$ is a symmetric, positive-definite matrix and K is a normalizing constant. It follows from (14) and (15) that

$$E(X_0 \mid X_1, \ldots, X_n) = -\sum_{k=1}^{n} \frac{d_{k0}}{d_{00}} X_k \quad \text{(a.s.)} \qquad (16)$$

so that in this case the conditional expectation is a *linear* function of the conditioning variables. In particular, for random variables X and Y with a bivariate normal distribution and means 0 (the case $n = 1$), (16) becomes

$$E(Y \mid X) = \frac{E(XY)}{E(X^2)} X \quad \text{(a.s.)} \qquad (17)$$

When $Y \in L_2$, the characterization (8) of conditional expectation as a projection shows that $E(Y \mid X_1, \ldots, X_n)$ is the *closest approximation* to Y in the L_2 norm by any random variable that is a function of the conditioning variables $X_1, \ldots, X_n$. In general, the best approximation by a *linear* function will not be as close as the conditional expectation; however, it is often much easier to evaluate and study. In the normally distributed case, the best linear approximation *is* the best approximation possible. Since normal (Gaussian) random variables and processes occur frequently in applications, this fact is of considerable importance.

---

[9] Normal distributions are discussed further in the postscript to Section 22.

5. WHEN DO RANDOM VARIABLES EXIST?

It is easy to give examples of probability spaces with interesting infinite families of random variables. For instance, take the space to be the unit interval [0, 1] with Lebesgue measure. For each $\omega \in [0, 1]$ and each positive integer k, let $X_k(\omega)$ equal the kth digit in the binary expansion[10] of ω. The $\{X_k\}$ then form a sequence of independent random variables, each taking the values 0 or 1 with probability $1/2$. In other words, this construction provides a model for an infinite sequence of tosses of a fair coin, based on the most familiar and concrete of measure spaces.

The problem here, however, is to reverse this process. Instead of starting with a space and functions (i.e., random variables), we want to begin by specifying the distributions the random variables will have, and then construct a probability space and functions that realize those distributions. In Chapters 2 and 3 there are many results of the form: "Let $X_1, X_2, \ldots$ be independent random variables with distribution functions $F_1, F_2, \ldots$. Then ... [some conclusion follows]." The question naturally arises as to whether, given the distribution functions $\{F_n\}$, there really exists a probability space $(\Omega, \mathcal{B}, P)$ on which such a family of random variables can be defined. If not, the theorems we will prove become less interesting. Fortunately, the answer is affirmative, and the number of distributions and random variables involved need not even be countable.

To get started, consider the case of a finite number n of distribution functions. Then we can take R^n with its field of Borel sets to serve as Ω and $\mathcal{B}$. The measure P is defined first for sets of a special form. If $I_k = (a_k, b_k]$ is a half-open interval for each k, the product set

$$S = \{(x_1, \ldots, x_n) \in R^n : a_k < x_k \leq b_k, \ k = 1, 2, \ldots, n\} \quad (1)$$

is called a *rectangle* and its measure is defined by

$$P(S) = \prod_{k=1}^{n} [F_k(b_k) - F_k(a_k)]. \quad (2)$$

The class of all finite unions of rectangles forms a field $\mathcal{F}$, and Theorem 1.1, the extension theorem, guarantees that the function P can be extended from $\mathcal{F}$ to the σ-field it generates—which in this case consists of the Borel

---

[10]The expansion is unique except for certain rationals, which of course form a set of measure zero. Any convention can be adopted for these, such as "use terminating binary expansions whenever they exist," or else the whole set can be excluded from the space.

sets in R^n. The result is a product measure on R^n such that the coordinate functions

$$X_k(x_1, \ldots, x_n) = x_k \tag{3}$$

have the desired distribution functions F_k. The details of this construction, including the verification that the function P defined by (2) for rectangles satisfies condition (1.3), can be found in Royden (1988).

What if we have an infinite sequence of distributions? Again we seek to construct a corresponding sequence of independent random variables, and the procedure is quite similar to the finite-dimensional case. For Ω, instead of R^n we choose the set R^∞ consisting of all sequences of reals, that is, all $\omega = (x_1, x_2, \ldots)$. A rectangle now means a set of ω restricted by a finite set of inequalities of the form (1), that is,

$$S = \{\omega \in R^\infty : a_k < x_k \leq b_k, \, k = 1, 2, \ldots, n\} \tag{4}$$

for any n and any constants $\{a_k \leq b_k\}$. The probability of any rectangle S is again defined by (2); thus $P(S)$ is the product of the desired probabilities that each coordinate belong to the appropriate interval, taken over those coordinates where the definition of S imposes a restriction. Once again, the collection of all finite unions of rectangles forms a field $\mathcal{F}$ of sets, and the extension theorem assures us that the function P can be extended from $\mathcal{F}$ to the σ-field it generates. The resulting measure is the product measure on R^∞, and again the coordinate functions defined by (3) provide the independent random variables we are seeking.

All these constructions are special cases of a more general theorem first stated and proved by Kolmogorov (see *Foundations*). In this theorem, the random variables need not be independent but can have any joint distributions whatsoever as long as these are consistent. The number of variables does not have to be countable.

To state the theorem and explain what "consistency" means, let us suppose that we are given, defined on some probability space, a family $\{X_t\}$ of real-valued random variables indexed by a parameter $t \in T$, where T is a subset of the real line.[11] That is, $\{X_t\}$ is a collection of functions of ω, where $t \in T$ and $\omega \in \Omega$; for each t, X_t is $\mathcal{B}$-measurable. Such an indexed family of random variables is called a *stochastic process*, and often (but

---

[11] Usually T will be either the entire real line, the integers, or the nonnegative part of one or the other of these. The theorem can be adapted to the more general situation where T is an arbitrary index set.

not always) the parameter t is thought of as representing time. For fixed $\omega = \omega_0$ the function $X_t(\omega_0)$ is called a *trajectory* or a *path function* of the process.

The joint probability distributions of finite subsets of these random variables are known as the *finite-dimensional distributions* of the process. More explicitly, let $t_1 < \cdots < t_n$ be numbers in T and suppose that S is any Borel set in R^n. Then the definition

$$P_{t_1,\ldots,t_n}(S) = P\left(\{\omega : [X_{t_1}(\omega), \ldots, X_{t_n}(\omega)] \in S\}\right) = P(\Phi^{-1}_{t_1,\ldots,t_n}(S)) \quad (5)$$

makes sense since the mapping $\Phi_{t_1,\ldots,t_n}$ from Ω into R^n, which sends ω into the point with coordinates $X_{t_1}(\omega), \ldots, X_{t_n}(\omega)$, is measurable. The set function $P_{t_1,\ldots,t_n}$ is a probability measure on R^n, and, of course, it is just the joint distribution of the random variables $X_{t_1}, \ldots, X_{t_n}$ as discussed in Section 3. Such a measure is defined for each ordered n-tuple of elements of T.

We will illustrate the consistency condition with a special case. Suppose that $r < s < t$ are members of T, and let τ be the projection from R^3 into R^2 that sends the point (x, y, z) into the point (x, z). Suppose that S is any Borel set in R^2; then $\tau^{-1}(S)$ consists of all the points in R^3 whose first and third coordinates define a point of R^2 belonging to S. Then clearly

$$\{\omega : [X_r(\omega), X_s(\omega), X_t(\omega)] \in \tau^{-1}(S)\} = \{\omega : [X_r(\omega), X_t(\omega)] \in S\},$$

and, therefore, we must have

$$P_{r,s,t}(\tau^{-1}(S)) = P_{r,t}(S). \quad (6)$$

A similar relation must hold between any two finite-dimensional distributions $P_\mathcal{M}$ and $P_\mathcal{N}$ whenever $\mathcal{M}$ and $\mathcal{N}$ are m and $(n + m)$-tuples of elements of T such that $\mathcal{M} \subset \mathcal{N}$. This is the meaning of "consistency" for the finite-dimensional distributions of the stochastic process.

Theorem 1 (Kolmogorov). *Suppose that for all ordered n-tuples $t_1 < \cdots < t_n$ of elements of T there is a probability measure $P_{t_1,\ldots,t_n}(S)$ defined on the Borel sets of R^n, and suppose also that these measures satisfy the general consistency condition described above. Then there exists a probability space $(\Omega, \mathcal{B}, P)$ and a family of random variables $\{X_t(\omega) : t \in T\}$ defined on it such that (5) holds; that is, the measures $P_{t_1,\ldots,t_n}(S)$ are the finite-dimensional distributions of the stochastic process $\{X_t\}$.*

We will not prove this theorem; the proof is given in Kolmogorov's *Foundations* and elsewhere, including Billingsley (1995). The idea is similar to the construction of product measures. The set Ω can be taken to be all the real-valued functions on T. The σ-field $\mathcal{B}$ obviously must include all the sets of the form $X_t^{-1}(S)$, where S is a Borel set in R^1 and $t \in T$, and Kolmogorov's construction selects for $\mathcal{B}$ the smallest σ-field containing these. Sets of functions that are defined by restrictions affecting at most a finite number of values of t are called *cylinder sets*; their probabilities can be defined from the given finite-dimensional distributions using (5). (The consistency conditions ensure that this definition is unambiguous.) Finally, the extension theorem is called upon to guarantee that this measure can be extended from the cylinder sets to the σ-field they generate, which is again the field $\mathcal{B}$.

Remark 1. Suppose we are seeking a family of independent random variables. Then the one-dimensional distributions are given for each t, and the other finite-dimensional distributions are products of these. In this situation the consistency condition is always satisfied.

Remark 2. Determining the finite-dimensional distributions is often considered equivalent to completely specifying a stochastic process. However, for some purposes the σ-field $\mathcal{B}$ of sets constructed in Kolmogorov's theorem is insufficiently rich. If $T = R^1$, for example, the set of all continuous functions does not belong to this field and so its probability is not determined by the finite-dimensional distributions.

We can demonstrate the last statement with the aid of an example. Suppose that $T = [0, 1]$ and assume that $X_t = 0$ a.s. for each t. Then all the finite-dimensional distributions are degenerate at the origin, and it follows that

$$P(X_t = 0 \text{ for all } t \in R) = 1 \tag{7}$$

for every countable set $R \subset T$. But that is *all* that follows! To see this, we define two probability measures on the set of all real functions on $[0, 1]$, say P_1 and P_2, that are both consistent with (7). P_1 will be a degenerate measure, giving measure one to the constant function $f(t) = 0$ for all t. To construct P_2, let ξ be a random variable with a uniform distribution on the unit interval and define

$$X_t = 0 \quad \text{for all } t \neq \xi; \qquad X_\xi = 1. \tag{8}$$

This defines another measure on the set of all real functions, one such that the set of functions that are 0 except at one point, and equal to 1 at that point, has unit mass. Thus,

$$P_1(X_t \text{ is continuous}) = 1; \qquad P_2(X_t \text{ is continuous}) = 0. \qquad (9)$$

However, both processes clearly satisfy (7) for every countable set R. It follows that the measure of the set of continuous functions is not determined by the finite-dimensional distributions; that is, this set does not belong to $\mathcal{B}$. Of course this phenomenon only occurs for noncountable families of random variables, and we will not meet it again until Chapter 4.

CHAPTER TWO

Laws of Large Numbers and Random Series

6. THE WEAK LAW OF LARGE NUMBERS

The oldest limit theorem of probability theory, due to James Bernoulli around 1700, can be stated in modern terms as follows: Let $X_1, X_2, \ldots$ be independent random variables such that each X_k takes only the values 1 or 0 with $P(X_k = 1) = p$ for each k, $0 < p < 1$. Then for any $\epsilon > 0$,

$$\lim_{n \to \infty} P\left(\left| \frac{X_1 + \cdots + X_n}{n} - p \right| > \epsilon \right) = 0. \tag{1}$$

The quantity $S_n = X_1 + \cdots + X_n$ is often interpreted as the number of "successes" in n independent "Bernoulli trials" with probability p for success.

Remark. All the random variables may be defined on one common probability space, but this is not necessary. For the statement and proof of Bernoulli's theorem, and of "weak" laws of large numbers in general, it is sufficient to know that $X_1, \ldots, X_n$ have a joint distribution for each n.

A proof of (1) that is direct and elementary can be based on the particular features of the *binomial distribution*

$$P(S_n = k) = \binom{n}{k} p^k (1-p)^{n-k}; \tag{2}$$

this is explained neatly in Feller (Vol. 1, 1968). However, simple methods of much greater generality were developed in the latter part of the nineteenth century by the Russian mathematician P. L. Chebyshev.[1] We will now describe Chebyshev's approach.

If X is any random variable, the quantity $m_k = E(X^k)$, when it exists, is called the kth *moment* of X; the first moment m_1 is simply the expectation. The second moment of $(X - m_1)$ is called the *variance* of X and usually denoted by σ^2; formally,

$$\text{var}(X) = \sigma^2 = E[\{X - E(X)\}^2]. \tag{3}$$

The square root σ of the variance is the *standard deviation* of X. It is easy to see that (i) if c is a constant, then $\text{var}(X + c) = \text{var}(X)$ and $\text{var}(cX) = c^2 \text{var}(X)$, (ii) $\text{var}(X) = 0$ if and only if X is constant a.s., and (iii) $\text{var}(X) = E(X^2) - E(X)^2 = m_2 - m_1^2$.

Problem 1. *Prove the preceding statements (i), (ii), and (iii).*

Problem 2. *Show that if m_k exists for some $k > 0$, then $E[(X - a)^k]$ exists for every real number a, and that also m_α exists for any α such that $0 \le \alpha < k$.*

It is possible to compute certain moments of a sum of independent random variables in terms of the moments of the individual terms, a tactic we will often find useful. This is especially easy for the first two moments. As noted in Section 2, the first moment is always a linear functional, with or without independence. As for second moments, we find in the case of independent random variables that *variances are additive;* that is, if X and Y are independent, then

$$\text{var}(X + Y) = \text{var}(X) + \text{var}(Y). \tag{4}$$

The proof is easy:

$$\text{var}(X + Y) = E[\{X - E(X) + Y - E(Y)\}^2]$$
$$= E[\{X - E(X)\}^2] + 2E[\{X - E(X)\}\{Y - E(Y)\}] + E[\{Y - E(Y)\}^2]$$
$$= \text{var}(X) + 0 + \text{var}(Y).$$

---

[1] At least five different transliterations of this name can be found in the literature.

(The cross product term in the second line vanishes because of Theorem 3.1 and the independence of X and Y.) Notice that (4) is a kind of Pythagorean theorem, since the random variables $X - E(X)$ and $Y - E(Y)$ can be thought of as orthogonal vectors in $L_2(\Omega, \mathcal{B}, P)$. The variances of X and Y are then the squared lengths of these vectors in the L_2 norm, and the variance of $(X + Y)$ represents, of course, the square on the hypotenuse.

Problem 3. *Show that the variance of the "random sum" (4.6) is*

$$\operatorname{var}(S) = E(M)\operatorname{var}(X) + E(X)^2 \operatorname{var}(M)$$

provided M and the X_k have finite second moments.

We will also need *Chebyshev's inequality,* theme and variations. Suppose that Y is any nonnegative random variable, and $a > 0$. Then

$$P(Y \geq a) \leq \frac{E(Y)}{a}. \tag{5}$$

The proof is very simple. Since $Y \geq 0$ its distribution is concentrated on $[0, \infty)$, and so from (2.8) we have

$$E(Y) = \int_0^\infty x\, dP_Y(x).$$

But then

$$E(Y) \geq \int_a^\infty x\, dP_Y(x) \geq \int_a^\infty a\, dP_Y(x) = aP(Y \geq a).$$

This inequality is most often encountered in the form

$$P(|X - E(X)| \geq b) \leq \frac{\operatorname{var}(X)}{b^2} \tag{6}$$

which holds for any random variable X with finite variance; (6) is obtained from (5) by choosing $Y = [X - E(X)]^2$ and $a = b^2$. Other useful variants of (5) are

$$P(|X| \geq b) \leq \frac{E(|X|^k)}{b^k}, \qquad \text{where } b > 0 \text{ and } k > 0, \tag{7}$$

and

$$P(X \geq b) \leq e^{-cb} E(e^{cX}), \qquad \text{where } c > 0. \tag{8}$$

Problem 4. *Derive (7) and (8) from (5).*

Problem 5. *Show that the bound in inequality (6) is sharp; that is, if $b > 0$ and $\sigma^2 > 0$ are given, there exists a random variable X for which equality occurs.*

It is easy, using these ideas, to prove a *weak law of large numbers* much more general than Bernoulli's.

Theorem 1. *Let $X_1, X_2, \ldots$ be a sequence of independent random variables with a common distribution having finite variance σ^2 and expected value μ. Then for any $\epsilon > 0$,*

$$\lim_{n \to \infty} P\left(\left|\frac{X_1 + \cdots + X_n}{n} - \mu\right| \geq \epsilon\right) = 0. \tag{9}$$

Proof. The mean of $S_n = X_1 + \cdots + X_n$ is $n\mu$ and the variance is $n\sigma^2$; thus the expectation of S_n/n is μ and its variance is σ^2/n. Applying Chebyshev's inequality (6) we have

$$P\left(\left|\frac{S_n}{n} - \mu\right| \geq \epsilon\right) \leq \frac{\sigma^2}{n\epsilon^2}, \tag{10}$$

which implies (9). □

Remark. Short as it is, the proof has two parts. By computing the mean and variance of S_n/n, we first show that this sequence of random variables converges to the constant μ in the mean of order 2. (This is strictly correct only when all the random variables are defined on a single probability space.) Then we use Chebyshev's inequality to make explicit a fact asserted in Problem 2.5—that convergence in the mean implies convergence in probability.

Remark. When setting down the foundations of probability, it is natural to introduce the concept of "expectation" since it is simply the integral of a function over a measure space, but it may not be so clear a priori that this integral has an intrinsic probabilistic meaning. Theorem 1 eliminates these doubts and shows that the expectation of a random variable, as the limit of S_n/n, is not only a mathematical convenience but an essential parameter in the theory.

THE WEAK LAW OF LARGE NUMBERS

Example. In the case of Bernoulli trials, $\mu = p$ and $\sigma^2 = pq$ ($q = 1 - p$) so that the estimate (10) becomes

$$P\left(\left|\frac{S_n}{n} - p\right| \geq \epsilon\right) \leq \frac{pq}{n\epsilon^2} \leq \frac{1}{4n\epsilon^2}. \tag{11}$$

Much sharper bounds can be obtained for this special situation, however.

The methods introduced above can yield a lot more. The distributions of the random variables $\{X_k\}$ enter only through their means and variances; let us call these μ_k and σ_k^2, respectively, and not assume that they are identical. The only function of independence is to ensure that the variances are additive; this in turn depends on the property that

$$E\{(X_i - \mu_i)(X_j - \mu_j)\} = 0 \quad \text{for } i \neq j. \tag{12}$$

Random variables satisfying (12) are said to be *uncorrelated*, a property much weaker than independence. [This is equivalent to saying that the "centered" variables $(X_k - \mu_k)$ are orthogonal in L_2.] The approach used for Theorem 1—calculating variances to show convergence in the mean and then applying Chebyshev's inequality—also serves to prove the following more general result.

Theorem 2. *Let $X_1, X_2, \ldots$ be uncorrelated random variables, and assume that*

$$\lim_{n \to \infty} \frac{1}{n^2} \sum_{k=1}^{n} \sigma_k^2 = 0. \tag{13}$$

Then for each $\epsilon > 0$,

$$\lim_{n \to \infty} P\left(\left|\frac{X_1 + \cdots + X_n}{n} - \frac{\mu_1 + \cdots + \mu_n}{n}\right| \geq \epsilon\right) = 0. \tag{14}$$

All this can be generalized even further by weakening the hypothesis that the random variables are uncorrelated; we indicate steps in this direction in Problems 7 and 8 below, which can again be solved by calculating variances. A little more terminology is in order first: The left side of (12) is called the *covariance* of X_i and X_j, and the covariance divided by the product of the two standard deviations is the *correlation coefficient*, usually denoted by one of the letters ρ or r. Formally, for random variables X and

Y with means μ_x and μ_y and standard deviations σ_x and σ_y, respectively, we define

$$\text{cov}(X,Y) = E\{(X - \mu_x)(Y - \mu_y)\} \quad \text{and} \quad \rho_{xy} = \frac{\text{cov}(X,Y)}{\sigma_x \sigma_y}. \quad (15)$$

(Of course, X and Y are "uncorrelated" when $\rho = 0$.) Given the random variables $X_1, \ldots, X_n$, their *covariance matrix* C is the $n \times n$ matrix with entries $c_{jk} = \text{cov}(X_j, X_k)$; the *correlation matrix* is defined analogously. If $E(X_k) = 0$ and the X_k are the elements of a column vector X, then $C = E(X X^T)$.[2] Since for any constants λ_j (forming the column vector λ) we have

$$0 \leq E\left(\left[\sum_{j=1}^{n} \lambda_j X_j\right]^2\right) = \sum_{j,k=1}^{n} c_{jk} \lambda_j \lambda_k = \lambda^T C \lambda,$$

the covariance matrix must always be nonnegative definite; it is (strictly) positive-definite unless the random variables X_k are linearly dependent. We will need this concept in Section 22.

Problem 6. *If random variables X and Y have finite, nonzero variances, show that their correlation coefficient ρ exists, that $-1 \leq \rho \leq 1$, that $\rho = 0$ if X and Y are independent, and that $\rho = \pm 1$ if and only if X and Y are linearly related.*

Problem 7. *Show that Theorem 2 still holds if the assumption that the random variables are uncorrelated is weakened by assuming only that their covariances are nonpositive.*

Problem 8. *Suppose that the correlation coefficient ρ_{ij} of X_i and X_j satisfies $\rho_{ij} \leq c(|j - i|)$ for some constants $c(k)$. Show that if*

$$[c(0) + \cdots + c(n-1)][\sigma_1^2 + \cdots + \sigma_n^2] = o(n^2)$$

as $n \to \infty$, then (14) is still true.

Example (Sampling From a Finite Population). Suppose that an urn contains N slips of paper carrying the numbers $x_1, \ldots, x_N$, and let X be the number on a slip choosen "at random" (i.e., with equal probabilities $1/N$)

---

[2] The notation A^T means the transpose of the matrix A.

from the urn. Without loss of generality we can assume that $E(X) = (\sum_{j=1}^{N} x_j)/N = 0$; then

$$\sigma^2 = \text{var}(X) = E(X^2) - E(X)^2 = \frac{1}{N} \sum_{j=1}^{N} x_j^2.$$

If $X_1, \ldots, X_n$ represent the outcomes of n random choices of slips from the urn *with replacement* of each slip chosen before the next draw, then the random variables X_k are independent and their sum $X_1 + \cdots + X_n = S_n$ has mean 0 and variance $n\sigma^2$.

Now suppose the slips are drawn randomly *without replacement*; of course, we must have $n \leq N$. In this situation each random variable X_k has the same distribution (and variance) as before, but they are no longer independent. It is plausible that the correlations are now negative since, for example, if X_j happens to take a large value, that value is no longer available as a possibility for X_k. In fact, it is not hard to show that

$$\text{cov}(X_j, X_k) = -\frac{\sigma^2}{N-1}. \tag{16}$$

The variance of the sum S_n is, of course, no longer $n\sigma^2$; it must be reduced by a "correction factor" depending on N and n:

$$\text{var}(S_n) = n\sigma^2 \frac{N-n}{N-1}. \tag{17}$$

Equation (17) is important in statistics since random sampling from a finite population is usually done without replacement.

Problem 9. *Derive* (16) *and* (17).

Another generalization of Theorem 1 lies in keeping the assumption that the random variables are independent and identically distributed (i.i.d.) but relaxing the requirement that they have finite variance. As long as the first moment exists, it turns out that (9) is still true, and more; as we will soon see, in the i.i.d. case S_n/n converges to μ not only in probability but almost surely. It is even possible for S_n/n to converge in probability, but not almost surely, in certain cases when the first moment does not exist.

The assumption that the variables have the same distribution can be modified (but not simply dropped) also in cases when their variances are infinite. We will not pursue these last two generalizations and merely remark that for independent random variables the problem of finding conditions for convergence in probability of S_n/n has been solved completely; see Gnedenko and Kolmogorov (1968).

One last comment: The term "weak" in the law of large numbers means that we are speaking of convergence in probability (and sometimes also of convergence in the mean). "Strong" laws are similar theorems where the conclusion is strengthened to convergence with probability one. We will examine this problem in detail very soon, but first pause to explore some consequences of the weak law in the classical (Bernoulli) situation.

7. THE WEIERSTRASS APPROXIMATION THEOREM

This section is a short digression; we will show how the law of large numbers in the special case of Bernoulli trials leads to an elegant method for approximating continuous functions by polynomials. Let $f(x)$ be any bounded function on $[0, 1]$, and let $X_1, X_2, \ldots, X_n$ be Bernoulli random variables; that is, they are independent and each X_k takes (only) the values 1 or 0 with probabilities p and $q = 1 - p$, respectively. Let $S_n = X_1 + \cdots + X_n$ be the number of successes in n trials. Then using the binomial distribution (6.2) we have

$$E\left[f\left(\frac{S_n}{n}\right)\right] = \sum_{k=0}^{n} f\left(\frac{k}{n}\right) \binom{n}{k} p^k(1-p)^{n-k} = B_n(p). \tag{1}$$

The right-hand side is called a *Bernstein polynomial* after the discoverer of this proof of Weierstrass' theorem; it is clearly a polynomial in p with degree at most n. When n is large, most of the weight of the binomial distribution is concentrated on values of k such that k/n is close to p; using this idea, we will prove that $B_n(p)$ approximates $f(p)$.

Theorem 1. *If f is continuous on the closed interval $[0, 1]$, then*

$$\lim_{n \to \infty} \max_{0 \le p \le 1} |B_n(p) - f(p)| = 0. \tag{2}$$

Proof. Since f is a continuous function on a compact space, it is bounded in absolute value (by M, say) and uniformly continuous. Thus, given $\epsilon > 0$, we can choose a positive δ such that for all x and y in $[0, 1]$, we have

$$|f(x) - f(y)| < \epsilon \quad \text{whenever } |x - y| < \delta. \tag{3}$$

THE WEIERSTRASS APPROXIMATION THEOREM

For any value of p, we can then write

$$|B_n(p) - f(p)| = \left| \sum_{k=0}^{n} \left[f\left(\frac{k}{n}\right) - f(p) \right] \binom{n}{k} p^k (1-p)^{n-k} \right|$$

$$\leq \sum_{k=0}^{n} \left| f\left(\frac{k}{n}\right) - f(p) \right| \binom{n}{k} p^k (1-p)^{n-k}. \quad (4)$$

Now divide the last sum in (4) into two parts, the first part of the sum containing only the terms where $|k/n - p| < \delta$ and the second part all the others. Because of (3) and the fact that the terms of the binomial distribution add to 1, the first sum must be less than ϵ. For the second sum, we have

$$\sum_{|\frac{k}{n}-p|\geq\delta} \left| f\left(\frac{k}{n}\right) - f(p) \right| \binom{n}{k} p^k (1-p)^{n-k}$$

$$\leq 2M \sum_{|\frac{k}{n}-p|\geq\delta} \binom{n}{k} p^k (1-p)^{n-k}$$

$$= 2M\, P\left(\left| \frac{S_n}{n} - p \right| \geq \delta \right) \leq \frac{2M}{4n\delta^2} \quad (5)$$

by the law of large numbers in the form (6.11). Therefore for large enough n the sum in (5) will be less than ϵ, and the entire sum in (4) is then less than 2ϵ. This completes the proof. □

An alternative approach requires stronger assumptions on f than mere continuity, but in exchange yields more information, namely estimates on the rate of convergence.

Theorem 2. *If f is continuously differentiable on $[0, 1]$, then*

$$\max_{0 \leq p \leq 1} |B_n(p) - f(p)| \leq \frac{C}{\sqrt{n}}, \quad (6)$$

while if f is twice continuously differentiable, we have

$$\max_{0 \leq p \leq 1} |B_n(p) - f(p)| \leq \frac{D}{n}. \quad (7)$$

Proof. Using definition (1), we have

$$B_n(p) - f(p) = E\left[f\left(\frac{S_n}{n}\right) - f(p)\right]. \tag{8}$$

If f is continuously differentiable, then by the mean value theorem

$$|f(x) - f(p)| \leq \max_{0 \leq u \leq 1} |f'(u)| \, |x - p| \tag{9}$$

for any $x \in [0, 1]$; combining (8) and (9) gives, for any p,

$$|B_n(p) - f(p)| \leq \max_{0 \leq u \leq 1} |f'(u)| E\left(\left|\frac{S_n}{n} - p\right|\right). \tag{10}$$

But from the Schwarz inequality (2.3) we have

$$E\left(\left|\frac{S_n}{n} - p\right|\right) \leq \left(E\left[\left(\frac{S_n}{n} - p\right)^2\right]\right)^{1/2} = \sqrt{\operatorname{var}\left(\frac{S_n}{n}\right)}. \tag{11}$$

That variance equals pq/n. Since $pq \leq 1/4$ for all $p \in [0, 1]$, (10) and (11) yield (6) with $C = \max |f'|/2$.

When the second derivative exists, (9) can be replaced by

$$f(x) - f(p) = f'(p)(x - p) + \frac{1}{2!}f''(\xi)(x - p)^2, \tag{12}$$

where ξ lies between p and x and hence within the interval $[0, 1]$. Then from (8) and (12),

$$|B_n(p) - f(p)| = \left|E\left(f'(p)\left(\frac{S_n}{n} - p\right) + \frac{1}{2!}f''(\xi)\left(\frac{S_n}{n} - p\right)^2\right)\right|$$

$$\leq \frac{1}{2!} \max_{0 \leq u \leq 1} |f''(u)| \operatorname{var}\left(\frac{S_n}{n}\right) \leq \frac{D}{n},$$

where $D = \max |f''|/8$. □

One might guess that additional smoothness conditions on f would produce even faster convergence, but this is not the case unless f is linear.

Problem 1. *Assuming that f has continuous derivatives of order 3 or more, show that*

$$\lim_{n\to\infty} n[B_n(p) - f(p)] = pq\frac{f''(p)}{2}.$$

[Hence provided f is not linear, so that $f''(p) \neq 0$ for some p, the convergence of $B_n(p)$ to $f(p)$ is still only at the rate $1/n$.] Is this result true assuming only two continuous derivatives?

8. THE STRONG LAW OF LARGE NUMBERS

We consider a situation almost the same as the setup for Theorem 6.1: $X_1, X_2, \ldots$ are again independent random variables with the same distribution, and at least the first moment of that distribution exists. The only change is that now we must assume that all the random variables are defined on the same probability space $(\Omega, \mathcal{B}, P)$.[3] Then the conclusion of Theorem 6.1—that the average of $X_1, \ldots, X_n$ tends to the common expected value—holds not only in the sense of convergence in probability or convergence in the mean (the latter when the second moments exist), but pointwise for almost all $\omega \in \Omega$ (i.e., for all ω except a set of probability 0). This result for the special case of Bernoulli trials was formulated and proved in 1909 by Emile Borel, who was among the first to apply the then novel ideas of Lebesgue measure and integration to probability.[4]

The proof of the *strong law of large numbers* under the conditions stated above, simply the existence of the expectation, will be completed in Section 9. We begin with something easier and assume that fourth moments are finite.

Theorem 1. *Let $X_1, X_2, \ldots$ be independent random variables with a common distribution that has mean μ, variance σ^2, and a finite fourth moment. Then*

$$P\left(\lim_{n\to\infty} \frac{X_1 + \cdots + X_n}{n} = \mu\right) = 1. \tag{1}$$

The simplest proof uses an easy lemma with a real-variables flavor.

---

[3] Recall that in Section 6 it was only necessary that $X_1, \ldots, X_n$ be defined on a common space for each n; in other words, that they have a joint distribution.

[4] "Les probabilités dénombrables et leur applications arithmétiques," *Rendiconti Circulo Mat. Palermo* 27 (1909), pp. 247–271.

Lemma 1. *Suppose that $Y_1, Y_2, \ldots$ are nonnegative random variables and that*

$$\sum_{n=1}^{\infty} E(Y_n) < \infty. \tag{2}$$

Then $Y_n \to 0$ with probability 1.

Proof of the Lemma. Let $Z_n = Y_1 + \cdots + Y_n$. Since the Y_n are nonnegative, the Z_n form an increasing sequence of functions; let $Z \leq \infty$ denote their limit (the sum of the Y_n). Using (2) and the monotone convergence theorem, we have

$$E(Z) = \lim_{n \to \infty} E(Z_n) = \lim_{n \to \infty} \sum_{k=1}^{n} E(Y_k) = \sum_{k=1}^{\infty} E(Y_k) < \infty$$

The function Z is therefore integrable, and so it is finite except on a set of measure 0. But Z is the sum of the Y_n, and the terms of a convergent series must tend to 0. Thus $Y_n \to 0$ a.s. as claimed.

Proof of the Theorem. Let $Y_n = (S_n - n\mu)^4/n^4$, where again $S_n = X_1 + \cdots + X_n$. To apply the lemma, we need to calculate $E(Y_n)$. If we expand the fourth power and then use independence and the multiplicative property of expectations (Theorem 3.1 and its corollary), it is easy to see that

$$E\left(\left[\sum_{k=1}^{n}(X_k - \mu)\right]^4\right) = n E([X_k - \mu]^4) + 6 \binom{n}{2} \sigma^4 \leq C n^2. \tag{3}$$

Thus

$$E(Y_n) = \frac{1}{n^4} E\left(\left[\sum_{k=1}^{n}(X_k - \mu)\right]^4\right) \leq \frac{C}{n^2}$$

so that (2) holds; therefore the lemma applies and $Y_n \to 0$ a.s. The conclusion of the theorem follows. $\square$

Remark. Using only the variance instead of the fourth moment, we would obtain a (divergent) harmonic series so this method of proof would not quite work. Nevertheless, we will soon see that only the first moment is necessary for the strong law (1) to hold.

THE STRONG LAW OF LARGE NUMBERS

Problem 1. *Suppose that $X_1, X_2, \ldots$ are independent and uniformly distributed on the interval $[0,1]$. [That is, they each have the density function $f(x) = 1$ for $0 \leq x \leq 1$; $f(x) = 0$ elsewhere.] Show that*

$$\lim_{n \to \infty} (X_1 X_2 \cdots X_n)^{1/n} = e^{-1} \quad \text{(a.s.)}$$

There is a simple but interesting application of the strong law that is also due to Borel. A real number x is called *normal in base d* if each digit occurs the "right" fraction of the time when x is expressed as a "decimal" (d-ecimal?) in the number system based on d. (That is, the fraction of the first n digits of the expansion having a particular value tends to $1/d$ as n tends to infinity.) The number x is then said to be *normal* if it is normal in base d for every $d > 1$. Rational numbers are never normal, although they may be normal in a particular base. (For example, in binary notation the rational number $1/3 = 0.010101\ldots$, so $1/3$ is normal in base 2.)

Theorem 2. *Almost all numbers are normal.*[5]

Proof. It is enough to prove this for the unit interval $[0,1]$. We can think of this interval, equipped with Lebesgue measure, as the probability space Ω. Now consider the binary expansion of an arbitrary point $x = 0.b_1 b_2 b_3 \ldots$, where each digit $b_k = 0$ or 1; adopt some convention to make the expansion unique. (For example, we could decree that no expansion can end with an infinite string of 1's.) Then each digit $b_k(x)$ is a random variable on the space; for example, $b_1(x) = 0$ when $x < 1/2$ and $= 1$ otherwise. It is not difficult to see that these variables are independent and that each of them takes the values 0 or 1 with equal probabilities $1/2$. By the strong law of large numbers, we have

$$P\left(\left\{x \in [0,1] : \lim_{n \to \infty} \frac{b_1(x) + \cdots + b_n(x)}{n} = \frac{1}{2}\right\}\right) = 1$$

so that almost all x are normal in base 2. A similar argument shows that almost all numbers x are normal in any other base d. Finally, the set of numbers in $[0,1]$ that are not normal is the union of the sets of numbers that are not normal in base d for $d = 2, 3, 4, \ldots$. Since each of these sets has measure 0 so does their union; the theorem is proved. Incidentally,

---
[5] That is, all except for a set of Lebesgue measure 0.

although a "random" number is almost surely normal, it is not so obvious how to exhibit a number whose normality can be established.[6] □

Problem 2. *Verify that the functions $\{b_k(x)\}$ are independent and that*

$$P(b_k(x) = 1) = \frac{1}{2}.$$

Problem 3. *Prove that almost all numbers are normal in base d for $d > 2$.*

Problem 4. *Let $X_1, X_2, \ldots$ represent the outcomes of general Bernoulli trials; that is, the X_k are independent and take only the values 1 or 0 with constant probabilities p and $q = 1 - p$, respectively. Define a random variable Z by the (convergent) series*

$$Z = \sum_{n=1}^{\infty} X_n 2^{-n}$$

and let $F_p(t) = P(Z \leq t)$ be the corresponding distribution function. Show that $F_{1/2}(t) = t$ for $0 \leq t \leq 1$, that F_p is continuous and strictly increasing for every value of $p \in (0, 1)$, and that if $p \neq 1/2$, then F_p is singular—that is, $F_p'(t) = 0$ for almost all (Lebesgue measure) values of t. [Hint: A continuous distribution function is singular if and only if the corresponding measure puts probability 1 on a set the Lebesgue measure of which is 0.]

9. THE STRONG LAW—CONTINUED

To improve Theorem 8.1 and obtain other almost sure results, some new tools are needed. One of the most basic is the following.

Lemma 1 (Borel and Cantelli). *Let $A_1, A_2, \ldots$ be events in a probability space, and let*

$$B = \limsup_{n \to \infty} A_n = \bigcap_{k=1}^{\infty} \bigcup_{n=k}^{\infty} A_n. \tag{1}$$

---

[6]Something analogous happens when we prove the existence of transcendental numbers by showing that the algebraic numbers are countable but the reals are not.

(i) If $\sum P(A_n) < \infty$, then $P(B) = 0$, and
(ii) if the events $\{A_n\}$ are independent and $\sum P(A_n) = \infty$, then $P(B) = 1$.

Proof. Since all measures are subadditive with respect to countable unions (disjoint or not), we have

$$P(B) \leq P\left(\bigcup_{n=k}^{\infty} A_n\right) \leq \sum_{n=k}^{\infty} P(A_n)$$

for each k. If $\sum P(A_n) < \infty$, the right-hand side tends to 0 as $k \to \infty$; this proves (i).

For the partial converse, it is enough to show that

$$P\left(\bigcup_{n=k}^{\infty} A_n\right) = 1 \qquad (2)$$

for each k since the intersection of a sequence of sets, each with probability 1, must also have probability 1. But for each $K > k$,

$$1 - P\left(\bigcup_{n=k}^{\infty} A_n\right) \leq 1 - P\left(\bigcup_{n=k}^{K} A_n\right)$$

$$= P\left(\bigcap_{n=k}^{K} [\Omega - A_n]\right) = \prod_{n=k}^{K} [1 - P(A_n)]$$

because the events A_n, and so also their complements, are independent. If $\sum P(A_n) = \infty$, the product tends to 0 as $K \to \infty$, and (2) follows. □

Remark 1. B is the set of all ω that belong to infinitely many of the sets A_n; we interpret $P(B) = 0$ as "only finitely many A_n occur" (a.s.), etc. Applications of part (ii) of the lemma are hampered by the requirement of independence, and several less restrictive sufficient conditions have been found. But the condition cannot be simply dropped—the case when all the A_n are the same, with probability neither 0 nor 1, is an extreme example.

Remark 2. The first part of the Borel–Cantelli lemma can be used instead of Lemma 8.1 to prove the version of the strong law of large numbers given in Section 8. To accomplish this, we combine the moment calculation (8.3) with Chebyshev's inequality in the form (6.7); the result is

$$P\left(\left|\sum_{k=1}^{n}(X_k - \mu)\right| \geq \epsilon n\right) \leq \frac{Cn^2}{(\epsilon n)^4}$$

for any $\epsilon > 0$. Since the sum of the right-hand side is finite, we see by the lemma that only finitely many of the inequalties $|\sum_{k=1}^{n}(X_k - \mu)| \geq \epsilon n$ can hold (with probability 1). Then taking a sequence of ϵ_n tending to 0, we can conclude that (8.1) also holds.

Problem 1. *Supply the details of this alternative proof of Theorem 8.1.*

Remark 3. The first part of Borel–Cantelli is actually a special case of Lemma 8.1. (Why?)

We also need a result that superficially resembles the variance form of Chebyshev's inequality (6.6), but which gives much more information about partial sums of independent random variables. The technique of the proof, as well as the inequality itself, will be useful later on.

Theorem 1 (Kolmogorov's Inequality). *Let $X_1, \ldots, X_n$ be independent random variables with means 0 and variances σ_k^2. Then for any $a > 0$,*

$$P\left(\max_{1 \leq k \leq n} |X_1 + \cdots + X_k| \geq a\right) \leq \frac{1}{a^2} \sum_{k=1}^{n} \sigma_k^2. \tag{3}$$

Proof. Let $S_k = X_1 + \cdots + X_k$ and define the events

$$A = \left\{\omega : \max_{1 \leq k \leq n} |S_k| \geq a\right\} \tag{4}$$

and

$$A_k = \{\omega : |S_j| < a \text{ for } j = 1, \ldots, k-1 \text{ and } |S_k| \geq a\}. \tag{5}$$

(Thus S_k is the *first* partial sum that equals or exceeds a.) Clearly the A_k are disjoint and $\bigcup_{k=1}^{n} A_k = A$. Since the means are 0, we have

$$\sum_{k=1}^{n} \sigma_k^2 = E(S_n^2) \geq E(S_n^2 \mathbf{1}_A) = \sum_{k=1}^{n} E(S_n^2 \mathbf{1}_{A_k}). \tag{6}$$

THE STRONG LAW—CONTINUED 47

[Again, $\mathbf{1}_A(\omega)$ denotes the "indicator function" of the set A; its value is 1 if $\omega \in A$, 0 if $\omega \notin A$.] Now

$$E(S_n^2 \mathbf{1}_{A_k}) = E([S_k + (S_n - S_k)]^2 \mathbf{1}_{A_k})$$
$$= E(S_k^2 \mathbf{1}_{A_k}) + 2E[S_k(S_n - S_k)\mathbf{1}_{A_k}] + E[(S_n - S_k)^2 \mathbf{1}_{A_k}]. \quad (7)$$

The middle term in the last line is 0. To see this, note that the random variables $(S_n - S_k)$ and $S_k \mathbf{1}_{A_k}$ are independent since they are functions of the disjoint blocks of variables $(X_{k+1}, \ldots, X_n)$ and $(X_1, \ldots, X_k)$, respectively;[7] we then use Theorem 3.1 and the fact that $E(S_n - S_k) = 0$. The last term in (7) is nonnegative. The first term is the integral of the function S_k^2 over the set A_k and by definition $|S_k| \geq a$ on this set. Putting all these things into (7) we obtain

$$E(S_n^2 \mathbf{1}_{A_k}) \geq a^2 P(A_k),$$

and substituting into (6), we have

$$\sum_{k=1}^n \sigma_k^2 \geq a^2 \sum_{k=1}^n P(A_k) = a^2 P(A)$$

which is the same as (3). $\square$

The next preliminary result is a consequence of Kolmogorov's inequality and the Borel–Cantelli lemma, but it deserves a label of its own.

Theorem 2. *Let $X_1, X_2, \ldots$ be independent random variables with means 0 and variances σ_k^2, and suppose that $\sum \sigma_k^2 < \infty$. Then*

$$P\left(\sum_{k=1}^\infty X_k \text{ converges}\right) = 1. \quad (8)$$

Proof. We will use the Cauchy criterion for convergence. Fix $\epsilon > 0$. By (3),

$$P\left(\max_{1 \leq k \leq m} |S_{n+k} - S_n| \geq \epsilon\right) \leq \frac{\sigma_{n+1}^2 + \cdots + \sigma_{n+m}^2}{\epsilon^2}$$

---

[7] The independence seems plausible, but it still needs proof. This missing link will be supplied in Section 11. That section does not depend on anything in Sections 6–10 and can, if desired, be read right away.

for any m. Since the sum of the variances converges, for any $\delta > 0$ we can choose n so that for all m the right-hand side is less than δ. It follows that

$$P(|S_i - S_j| \geq 2\epsilon \text{ for some } i \geq n, \, j \geq n) \leq \delta.$$

Consequently, for the ϵ we have chosen,

$$P(|S_i - S_j| \geq 2\epsilon \text{ for arbitrarily large } i, j) = 0. \tag{9}$$

Now choose a sequence $\{\epsilon_n\}$ with limit 0; (9) must hold simultaneously for all the ϵ_n so that finally we have

$$P(\exists\, \epsilon > 0 : |S_i - S_j| \geq \epsilon \text{ for arbitrarily large } i, j) = 0.$$

This says that the partial sums $\{S_n\}$ form a Cauchy sequence with probability 1, and the convergence (8) follows. $\square$

Remark. The assumptions of the theorem state that the random variables $\{X_n\}$ are orthogonal in the space $L_2(\Omega)$ and that their sum is convergent in the L_2 norm. For many specific orthogonal series, it can be shown that such L_2 convergence implies pointwise convergence almost everywhere. This is true in particular for ordinary Fourier series, but that is a difficult theorem! In our case, the added assumption of independence, stronger than orthogonality, makes it much easier to obtain this conclusion.

An elementary fact about limits is now necessary.

Lemma 2. *If $\{a_n\}$ is a sequence of numbers such that $\sum(a_n/n)$ converges, then $(a_1 + \cdots + a_n)/n \to 0$.*

Problem 2. *Prove this lemma.*

Combining this with Theorem 2 yields a big improvement on Theorem 8.1:

Theorem 3. *Let $X_1, X_2, \ldots$ be independent random variables with means μ_n and variances σ_n^2. Suppose that $\sum \sigma_n^2/n^2 < \infty$. Then*

$$P\left(\lim_{n \to \infty} \frac{1}{n} \sum_{k=1}^{n} (X_k - \mu_k) = 0\right) = 1. \tag{10}$$

THE STRONG LAW—CONTINUED

Proof. The random variables $(X_n - \mu_n)/n$ have means 0 and variances σ_n^2/n^2 so that Theorem 2 applies; the result is

$$P\left(\sum_{n=1}^{\infty} \frac{X_n - \mu_n}{n} \text{ converges}\right) = 1.$$

Applying Lemma 2 to this conclusion, we obtain (10). Note that, in particular, if the X_n are identically distributed then the existence of the second moment (rather than the fourth) is sufficient for the strong law to hold. □

Two more facts about distribution functions will be needed.

Lemma 3. *Let F be a probability distribution function. Then*

$$\int_0^{\infty} x\, dF(x) = \int_0^{\infty} (1 - F(x))\, dx$$

and

$$\int_{-\infty}^0 x\, dF(x) = -\int_{-\infty}^0 F(x)\, dx, \tag{11}$$

where in both cases the two sides exist or diverge together.

Problem 3. *Prove Lemma 3. [Hint: Formally, both parts of (11) are easily obtained using integration by parts. Showing the convergence if and only if, however, requires a little more thought.]*

Lemma 4. *If F is a distribution function such that $\int_{-\infty}^{\infty} |x|\, dF(x) < \infty$, then*

$$\sum_{n=1}^{\infty} \frac{1}{n^2} \int_{-n}^{n} x^2\, dF(x) < \infty. \tag{12}$$

Proof. To obtain (12), let

$$a_{n+1} = \int_n^{n+1} x\, dF(x) + \int_{-(n+1)}^{-n} |x|\, dF(x);$$

then evidently,

$$a_n \geq 0 \quad \text{and} \quad \sum_{n=1}^{\infty} a_n = \int_{-\infty}^{\infty} |x|\, dF(x) < \infty.$$

But it is easy to see that

$$\int_{-(n+1)}^{-n} x^2 \, dF(x) + \int_{n}^{n+1} x^2 \, dF(x) \leq (n+1) a_{n+1},$$

and so

$$\sum_{n=1}^{\infty} \frac{1}{n^2} \int_{-n}^{n} x^2 \, dF(x) \leq \sum_{n=1}^{\infty} \frac{1}{n^2} \sum_{k=1}^{n} k a_k = \sum_{k=1}^{\infty} k a_k \sum_{n=k}^{\infty} \frac{1}{n^2}.$$

The inner sum (over n) in the last expression above is asymptotic to $1/k$ for large k, and so the terms of the summation over k are asymptotic to a_k. Since by assumption $\sum a_k < \infty$, (12) follows. □

At last we are ready for this section's main result, the definitive form of the strong law of large numbers for identically distributed variables.

Theorem 4. *Let $X_1, X_2, \ldots$ be independent random variables having the same probability distribution; assume that their expected value μ exists. Then*

$$P\left(\lim_{n \to \infty} \frac{X_1 + \cdots + X_n}{n} = \mu\right) = 1. \qquad (13)$$

If the expected value does not exist, then

$$P\left(\limsup_{n \to \infty} \frac{|X_1 + \cdots + X_n|}{n} = +\infty\right) = 1. \qquad (14)$$

Proof. We can assume that $\mu = 0$, for if not we can work with the random variables $X_n - \mu$ instead of the X_n. All the techniques developed so far depend on the existence of higher moments, which is no longer assumed. But these techniques *can* be applied to the "truncated" variables defined by

$$Y_n = \begin{cases} X_n & \text{if } |X_n| \leq n \\ 0 & \text{otherwise.} \end{cases} \qquad (15)$$

Define also variables Z_n so that $X_n = Y_n + Z_n$ for each n. The idea of the proof is to show that all but a finite number of the Z_n are 0 and therefore they do not affect the limiting behavior of the partial sums. The Y_n, on the other hand, have finite variances so that Theorem 3 can be applied.

THE STRONG LAW—CONTINUED

First we dispose of the Z_n. Note that

$$P(Z_n \neq 0) = P(|X_n| > n) \leq 1 - F(n) + F(-n),$$

where F is the common distribution function of the X_n. (Inequality occurs only when there is a positive probability that $X_n = -n$.) But then

$$\sum_{n=1}^{\infty} P(Z_n \neq 0) \leq \sum_{n=1}^{\infty} [1 - F(n) + F(-n)]$$

$$\leq \int_0^{\infty} [1 - F(x)] \, dx + \int_{-\infty}^0 F(x) \, dx$$

$$= \int_{-\infty}^{\infty} |x| \, dF(x) = E(|X_n|) < \infty,$$

using Lemma 3 and the assumption that the mean exists. Thus by the Borel–Cantelli lemma

$$P(Z_n \neq 0 \text{ for infinitely many } n) = 0. \tag{16}$$

Next we will apply Theorem 2 to the variables Y_n. We have

$$\text{var}(Y_n) \leq E(Y_n^2) = \int_{-n}^{n} x^2 \, dF(x),$$

and so by Lemma 4

$$\sum_{n=1}^{\infty} \frac{\text{var}(Y_n)}{n^2} < \infty.$$

We conclude that (10) holds for the Y_n. Because of the truncation the means $\mu_n = E(Y_n)$ may not be exactly 0, but they do tend to 0 since

$$\lim_{n \to \infty} E(Y_n) = \lim_{n \to \infty} \int_{-n}^{n} x \, dF(x) = \int_{-\infty}^{\infty} x \, dF(x) = 0 \tag{17}$$

by the dominated convergence theorem. If we combine (16), (10), and (17), it finally is clear that

$$P\left(\lim_{n\to\infty} \frac{X_1 + \cdots + X_n}{n} = 0\right) \tag{18}$$

$$= P\left(\frac{Y_1 - \mu_1 + \cdots + Y_n - \mu_n}{n} + \frac{\mu_1 + \cdots + \mu_n}{n} + \frac{Z_1 + \cdots + Z_n}{n} \to 0\right) = 1,$$

which proves the main part of the theorem.

To prove the converse, we assume that $E(|X_n|) = \infty$. For any constant $C > 0$, define the events

$$A_n = \{\omega : |X_n(\omega)| \geq Cn\}.$$

From Lemma 3 it is easy to see that $\sum P(A_n) = \infty$. The A_n are independent, so by the second part of the Borel–Cantelli lemma we have

$$P(|X_n| \geq Cn \text{ for arbitrarily large } n) = 1 \tag{19}$$

for each C. From this we can see that (14) holds, which completes the proof of the theorem. □

Problem 4. *Supply the missing steps in the proof of (19).*

Problem 5. *Suppose that $X_1, X_2, \ldots$ are independent, identically distributed, positive random variables. Establish the necessary and sufficient condition for the existence with probability 1 of*

$$gm = \lim_{n\to\infty} (X_1 X_2 \cdots X_n)^{1/n},$$

and find the value of gm. [Recall Problem 8.1.]

10. CONVERGENCE OF RANDOM SERIES

In the last section we obtained one result on random series (Theorem 9.2); here we will take the matter further and establish a necessary and sufficient condition for the almost sure convergence of a series of independent random variables. For any $c > 0$, we define the *truncation at c* of a random variable X by

$$X^{(c)} = \begin{cases} X & \text{if } |X| \leq c; \\ 0 & \text{otherwise.} \end{cases} \tag{1}$$

The following result is known as the "Three-series theorem."

Theorem 1 (Kolmogorov). *Let $X_1, X_2, \ldots$ be independent random variables. Then*

$$P\left(\sum_{n=1}^{\infty} X_n \text{ exists}\right) = 1 \qquad (2)$$

if and only if, for some $c > 0$, each of the following series converges:

$$\sum_{n=1}^{\infty} P(|X_n| > c), \quad \sum_{n=1}^{\infty} E(X_n^{(c)}), \quad \sum_{n=1}^{\infty} \text{var}(X_n^{(c)}). \qquad (3)$$

Proof. Let us first assume the convergence of the three series. The first one together with the Borel–Cantelli lemma implies that $X_n^{(c)} = X_n$ for all except finitely many n (a.s.), so the problem is reduced to showing the convergence of $\sum X_n^{(c)}$. Because their variances converge, we know from Theorem 9.2 that

$$P\left(\sum_{n=1}^{\infty} [X_n^{(c)} - E(X_n^{(c)})] \text{ converges}\right) = 1, \qquad (4)$$

and since by assumption the sum of the expected values in (4) converges, we conclude that $\sum X_n^{(c)}$ converges a.s. and so that (2) holds.

The converse—proving the necessity of (3)—takes a little longer because we have not made all the necessary preparations in advance. Assume (2). Then $P(X_n \to 0) = 1$ so that only finitely many of the independent events $|X_n| > c$ can occur. By part (ii) of Borel–Cantelli, this implies that the first series in (3) must be convergent. It also implies that the series $\sum X_n^{(c)}$, which differs from the series $\sum X_n$ in only a finite number of terms, is itself convergent a.s.

To show convergence of the other two parts of (3), we need two lemmas. The first of these gives a lower bound corresponding to the Kolmogorov inequality's upper bound.

Lemma 1. *Let $X_1, \ldots, X_n$ be independent random variables with means 0 and variances $\sigma_1^2, \ldots, \sigma_n^2$, and assume that $|X_k| \leq c$ a.s. for each $k \leq n$. Then for any $a > 0$,*

$$P\left(\max_{k \leq n} |X_1 + \cdots + X_k| \geq a\right) \geq 1 - \frac{(a+c)^2}{\sum_{k=1}^{n} \sigma_k^2}. \qquad (5)$$

Proof. The approach is very similar to the proof of Theorem 9.1, and we will define the sets A and A_k exactly as in (9.4) and (9.5). Once again, we use the decomposition

$$E(S_n^2 \mathbf{1}_A) = \sum_{k=1}^{n} E(S_n^2 \mathbf{1}_{A_k})$$

and rewrite a typical term of the sum just as in (9.7). As before, the middle (cross product) term equals 0. This time we are seeking an upper bound and so we do not discard the last term; instead we note that, like the middle one, it is the expectation of a product of two independent factors and so

$$E([S_n - S_k]^2 \mathbf{1}_{A_k}) = E([X_{k+1} + \cdots + X_n]^2) E(\mathbf{1}_{A_k}) = \sum_{j=k+1}^{n} \sigma_j^2 P(A_k).$$

To bound the first term in (9.7) from above, notice that when $\omega \in A_k$ we must have $|S_{k-1}| < a$, and so $|S_k| < (a+c)$ since $|X_k|$ is bounded by c. Thus,

$$E(S_k^2 \mathbf{1}_{A_k}) \le (a+c)^2 P(A_k).$$

Putting all these pieces into the right-hand side of (9.7), we find that

$$E(S_n^2 \mathbf{1}_{A_k}) \le P(A_k) \left((a+c)^2 + \sum_{j=k+1}^{n} \sigma_j^2 \right) \le P(A_k) \left((a+c)^2 + \sum_{j=1}^{n} \sigma_j^2 \right),$$

and summing produces

$$E(S_n^2 \mathbf{1}_A) \le \left((a+c)^2 + \sum_{k=1}^{n} \sigma_k^2 \right) P(A). \tag{6}$$

But on the complement of A we must have $|S_n| < a$, and so

$$E(S_n^2 \mathbf{1}_A) = E(S_n^2) - E(S_n^2 \mathbf{1}_{\Omega - A}) \ge \sum_{k=1}^{n} \sigma_k^2 - a^2 P(\Omega - A)$$

$$= \sum_{k=1}^{n} \sigma_k^2 - a^2 + a^2 P(A). \tag{7}$$

Combining (6) and (7) and solving for $P(A)$, we obtain

$$P(A) \geq \frac{\sum_{k=1}^{n} \sigma_k^2 - a^2}{(a+c)^2 - a^2 + \sum_{k=1}^{n} \sigma_k^2} = 1 - \frac{(a+c)^2}{(a+c)^2 - a^2 + \sum_{k=1}^{n} \sigma_k^2}$$

from which (5) immediately follows. □

Lemma 2. *Let $X_1, X_2, \ldots$ be independent random variables with means 0 and variances σ_n^2, and assume that for some $c > 0$, $|X_n| \leq c$ a.s. for each n. Then if the series $\sum X_n$ converges a.s., the series $\sum \sigma_n^2$ must converge as well.*

Proof. From the almost sure convergence of the series $\sum X_n$, it follows by the Cauchy criterion that

$$\lim_{n \to \infty} \max_{m \geq 0} |S_{n+m} - S_n| = 0 \quad \text{(a.s.)},$$

and so that maximum also converges to 0 in probability. Thus we have

$$\lim_{n \to \infty} P\left(\max_{m \geq 0} |S_{n+m} - S_n| \geq a\right) = 0 \quad \text{for any } a > 0.$$

Fix an n for which this probability is, say, $< 1/2$. By Lemma 1,

$$P\left(\max_{0 \leq m \leq N} |S_{n+m} - S_n| \geq a\right) \geq 1 - \frac{(a+c)^2}{\sigma_{n+1}^2 + \cdots + \sigma_{n+N}^2}.$$

If the sum of the variances diverges, this would mean that the left-hand side tends to 1 as $N \to \infty$. This contradiction proves the lemma. □

Remark. Lemma 2 is another example of the power of independence, for if the random variables X_n were merely orthogonal and uniformly bounded, the almost sure convergence of $\sum X_n$ would *not* imply convergence in the L_2 norm.

Let us return to the proof of Theorem 1. Our intention, naturally, is to use Lemma 2, but since the means of the X_n^c are not 0, the lemma does not apply directly. This difficulty can be overcome by a trick of symmetrization. Suppose that $Y_1, Y_2, \ldots$ is another sequence of independent random variables that are independent of the X_n and such that Y_j has the same probability distribution as X_j for each j.[8] Now consider the random

---

[8] It is not necessary that the X_n and Y_n all be defined on the original probability space. One way to construct these variables is to use a new probability space that is the direct product of the original space Ω with itself.

series $\sum Z_n$, where

$$Z_n = X_n^c - Y_n^c. \qquad (8)$$

Since both $\sum X_n^c$ and $\sum Y_n^c$ converge almost surely, so does $\sum Z_n$. The Z_n are uniformly bounded (by $2c$) and clearly by symmetry $E(Z_n) = 0$. Thus Lemma 2 *does* apply to $\sum Z_n$ and states that the sum of the variances of the Z_n must be convergent. But $\text{var}(Z_n) = 2\text{var}(X_n^c)$; hence the third series in Eq. (3) must converge as well.

Finally, because of Theorem 9.2 the series $\sum [X_n^c - E(X_n^c)]$ is convergent with probability 1. But since $\sum X_n^c$ converges a.s., the convergence of $\sum E(X_n^c)$, which is the only part of (3) still missing, follows immediately. $\square$

Remark 1. As the proof shows, if (3) holds for some $c > 0$ it has to hold for any such c, and in this case the random series $\sum X_n$ converges almost surely. As we will see in the next section, if the probability of convergence is not 1, it must be 0; there can be no middle ground.

Remark 2. For any series of independent random variables, the probability of absolute convergence is also either 0 or 1 by the above remark applied to $\sum |X_n|$; it is perfectly possible for such a series to converge conditionally with probability 1. But for random series there is an intermediate concept between absolute and conditional convergence: The series may converge conditionally and yet have the property, which would be implied by absolute convergence, that it remains convergent after any rearrangement of the order of its terms. This possibility is illustrated below.

Example. Let $X_1, X_2, \ldots$ be independent "coin-tossing" random variables; that is, each variable takes the values $+1$ and -1 with probabilities $1/2$. Consider the random series

$$\sum_{n=1}^{\infty} Z_n = \sum_{n=1}^{\infty} \left(\frac{X_n}{n} + \mu_n \right), \qquad (9)$$

where the μ_n are constants tending to 0. It is not hard to see that the probability of absolute convergence is always 0. In condition (3) of Theorem 1 the first and last series clearly must converge, so the convergence of (9) hinges on that of $\sum E(Z_n) = \sum \mu_n$. If *this* series is absolutely convergent, therefore, the random series will converge (a.s.) no matter how the terms are reordered, but if $\sum \mu_n$ is conditionally convergent, any rearrangement that destroys its convergence will do the same for the convergence of the random series (9). General criteria exist for these modes of convergence, and for the convergence of $\sum(X_n - a_n)$ for some sequence of "centering" constants a_n, but we will not pursue the subject further here. [See Gnedenko and Kolmogorov (1968) or Loeve (1977).]

Problem 1. *Show that with probability 1 the series (9) is not absolutely convergent.*

Problem 2. *Let $\sum Z_n = \sum X_n/n^\alpha$, where again the X_n are coin-tossing variables and $\alpha > 0$ is a constant. Show that $\sum Z_n$ converges almost surely if and only if $\alpha > 1/2$.*

Problem 3. *Let $\sum Z_n = \sum c_n X_n$ be a general series with* random *signs; that is, $c_n \geq 0$ are constants and the X_n are independent random variables taking the values $+1$ and -1 with probabilities p and $1 - p$. Find the conditions under which the series converges a.s., and also the conditions under which the convergence is unaltered by any rearrangement of the terms. [Hint: The answer depends on whether or not $p = 1/2$.]*

11. MORE ON INDEPENDENCE; THE 0–1 LAW

Kolmogorov's *0–1 law* states roughly that if $X_1, X_2, \ldots$ is a sequence of independent random variables and if A is an event defined in terms of these X_n that is invariant under changes to any finite number of them, then either $P(A) = 1$ or $P(A) = 0$. For example, a sequence or a series of independent r.v.s must converge or diverge with probability 1; there is no middle ground. The proof of the 0–1 law itself is very short, but we must first make some preparations that are also needed in order to clarify certain points about independence that have been treated rather casually so far.

Any set X of random variables on a probability space $(\Omega, \mathcal{B}, P)$ determines a Borel field $\mathcal{B}(X)$ defined as the smallest sub σ-field of $\mathcal{B}$ with respect to which each $X \in X$ is measurable. In fact, $\mathcal{B}(X)$ is the intersection of all the Borel subfields of $\mathcal{B}$ that contain every set of the form $\{\omega : X(\omega) \in S\}$ where $X \in X$ and S is a Borel set of real numbers. We then say that $\mathcal{B}(X)$ is the σ-field *generated* by X. If X consists of a single random variable X, then $\mathcal{B}(X)$ or $\mathcal{B}(X)$ consists of the inverse images under X of all the Borel sets in the real line.[9]

---

[9] This paragraph recapitulates matters discussed in Chapter 1, especially Section 4. Since the reader may have postponed reading that section, however, it seemed desirable to repeat them here.

Suppose now that $X_1, X_2, \ldots$ are independent random variables and that $A_k \in \mathcal{B}(X_k)$ for each k; then by the definition of independence

$$P(A_1 \cap A_2 \cap \cdots \cap A_n) = \prod_{k=1}^{n} P(A_k) \qquad (1)$$

for any finite n. It is useful to generalize this a bit: If $\mathcal{B}_1, \mathcal{B}_2, \ldots$ is a sequence of σ-fields (subfields of $\mathcal{B}$) such that (1) holds whenever $A_k \in \mathcal{B}_k$, we will say that the σ-fields $\mathcal{B}_k$ are independent. We need to know that any set in the field generated by some subset of the $\mathcal{B}_k$ is also independent of sets in all the other fields in the sequence. More precisely, we will prove the following.

Theorem 1. *Let $\mathcal{B}_k, -\infty < k < \infty$, be independent Borel fields as defined above, and let $\mathcal{G}$ be the σ-field generated by any subset (finite or not) of the fields $\mathcal{B}_k$ with $k \leq 0$. Then the fields $\mathcal{G}, \mathcal{B}_1, \mathcal{B}_2, \ldots$ are independent.*

Proof. We may as well assume that $\mathcal{G}$ is the field generated by all the fields $\mathcal{B}_k$ with $k \leq 0$; let G denote any set in this field. Let A denote a fixed set of the form

$$A = A_1 \cap A_2 \cap \cdots \cap A_n \qquad (2)$$

where $A_k \in \mathcal{B}_k$ and each $k > 0$. Then we must show that

$$P(A \cap G) = P(A)P(G). \qquad (3)$$

We can assume that $P(A) > 0$ since otherwise (3) is trivial.

To begin, we note that if the set G is of the form

$$G = A_0 \cap A_{-1} \cap \cdots \cap A_{-m} \qquad (4)$$

where $A_j \in \mathcal{B}_j$ with each $j \leq 0$, then (3) holds because of the assumption of the theorem that the fields $\mathcal{B}_n$ are independent. The class $\mathcal{F}$ of all finite unions of sets of the form (4) forms a finitely additive field, and we will next show that (3) holds for every set $G \in \mathcal{F}$. Suppose that

$$G = G_1 \cup G_2 \cup \cdots \cup G_r, \qquad (5)$$

where each G_k is of the form (4), represents any such set. To verify (3) for this G, we can use the "inclusion–exclusion" formula for the probability of a finite union of sets:

$$P(A \cap G) = P\left(A \cap \left[\bigcup_{k=1}^{r} G_k\right]\right) = P\left(\bigcup_{k=1}^{r} [A \cap G_k]\right) \tag{6}$$
$$= \sum_{i} P(A \cap G_i) - \sum_{i<j} P(A \cap G_i \cap G_j) + \sum_{i<j<k} P(A \cap G_i \cap G_j \cap G_k) - \cdots;$$

the series has r terms. But each set of the form $G_i \cap G_j \cap G_k$, etc., that appears in (6) is the kind of G represented in (4), for which (3) is already known to hold. Thus for each term in any of the sums in the last line of (6), the set A can be removed from the intersection and the factor $P(A)$ inserted. The result is that every term in the inclusion–exclusion sum has a factor $P(A)$. When this term is factored out, the sum that remains equals $P(\bigcup_{k=1}^{r} G_k) = P(G)$, again by the inclusion–exclusion formula. Equation (6) thus reduces to (3), which we now see must hold for every $G \in \mathcal{F}$.

The rest is easy. Let P' denote the conditional probability measure given A; that is, define $P'(B) = P(B \cap A)/P(A)$ for any set B. Then both P' and P are probability measures on $\mathcal{B}$, and, as we have just shown, the two agree on all the sets of the field $\mathcal{F}$. By the uniqueness part of the basic extension theorem for measures (Theorem 1.1), we conclude that P' and P must coincide at least on the smallest Borel field containing $\mathcal{F}$. But this smallest field is exactly the field $\mathcal{G}$ generated by the $\mathcal{B}_k$ with $k \leq 0$. Hence (3) holds for all sets $G \in \mathcal{G}$, which completes the proof. □

Corollary 1. *Under the conditions of the theorem, if $\mathcal{G}_1$ and $\mathcal{G}_2$ are the Borel fields generated by two disjoint collections of the fields $\mathcal{B}_k$, then $\mathcal{G}_1$ and $\mathcal{G}_2$ are independent.*

Proof. Suppose that the two collections are taken from among the $\mathcal{B}_k$ with $k \leq 0$ and those with $k > 0$, respectively. By the theorem, $\mathcal{G}_1$ together with the $\mathcal{B}_k$ for $k > 0$ forms a sequence of independent fields. Another application of the theorem then finishes the job. □

Corollary 2. *Let $X_1, \ldots, X_{n+m}$ be independent random variables, and let f and g be real-valued Borel functions on R^n and R^m, respectively. Then the random variables*

$$Y = f(X_1, \ldots, X_n) \quad \text{and} \quad Z = g(X_{n+1}, \ldots, X_{n+m})$$

are independent.

Remark. The second corollary is a direct application of the first one. Note that it justifies some of our earlier manipulations of allegedly independent random variables, such as in the proof of Kolmogorov's inequality. [See the discussion following Eq. (9.7) and the footnote there.] It also validates the simple inductive approach we could not use legitimately back in Section 3 when we wanted to extend Theorem 3.1 (the multiplicative property of the expected value) from two to several independent random variables. Corollary 2 implies that if X, Y, and Z are independent, then so are X and (YZ), etc., so the induction can now be carried out.

We turn to the famous 0–1 law. If $X_1, X_2, \ldots$ is any sequence of random variables, the σ-fields

$$\mathcal{M}_n = \mathcal{B}(\{X_n, X_{n+1}, \ldots\}) \tag{7}$$

are obviously decreasing as n increases;[10] their intersection is called the *tail field* of the sequence. (The tail field contains at least the sets Ω and $\emptyset$.)

Theorem 2. *If A is any set belonging to the tail field of a sequence of independent random variables, then either $P(A) = 0$ or $P(A) = 1$.*

Proof. Let $\mathcal{M}_\infty$ denote the tail field. By Theorem 1 we know that for any n the fields $\mathcal{B}(X_1), \ldots, \mathcal{B}(X_{n-1}), \mathcal{M}_n$ are independent. But $\mathcal{M}_\infty \subset \mathcal{M}_n$ for all n, and it follows that all the fields

$$\mathcal{B}(X_1), \ \mathcal{B}(X_2), \ldots, \ \mathcal{M}_\infty$$

are independent. Again applying Theorem 1, we see that the tail field $\mathcal{M}_\infty$ is independent of the field $\mathcal{M}_1$ generated by all the random variables $\{X_n\}$. But this field clearly contains the tail field, and therefore the latter must be independent of itself! This means that any two sets in $\mathcal{M}_\infty$ (distinct or not) are independent, and in particular if A is such a set we must have

$$P(A) = P(A \cap A) = P(A)P(A).$$

Consequently, $P(A)$ must be either 0 or 1. $\square$

---

[10] More precisely, the $\mathcal{M}_n$ are nonincreasing.

Remarks. A Borel field of sets all of which have probability 0 or 1, such as the tail field of a sequence of independent random variables, is equivalent to the trivial field $\{\Omega, \varnothing\}$ if sets that differ only by a set of measure 0 are identified. Any random variable measurable with respect to such a field must be constant almost surely.

Problem 1. *Justify the assertions at the beginning of this section by showing that, for any sequence of random variables, the sets*

$$L_1 = \left\{\omega : \lim_{n\to\infty} X_n \text{ exists}\right\} \quad \text{and} \quad L_2 = \left\{\omega : \sum_{n=1}^{\infty} X_n \text{ converges}\right\}$$

both belong to the tail field of the sequence.

Problem 2. *Suppose that $X_1, X_2, \ldots$ are independent random variables. Show that the radius of convergence R of the random power series*

$$S(t) = \sum_{n=1}^{\infty} X_n t^n \tag{8}$$

is constant almost surely.

Problem 3. *Suppose that the coefficients X_k of the power series (8) are identically distributed as well as independent, and that $P(X_k = 0) < 1$. Show that the radius of convergence R must either be almost surely 0 or almost surely 1, and that $R = 1$ if and only if*

$$E[\log_+(|X_k|)] < \infty, \tag{9}$$

where $\log_+(s) = \log(s)$ if $s \geq 1$, and $\log_+(s) = 0$ for $s < 1$. [Hint: First show that $R \leq 1$ a.s. Then suppose that $R \geq p$ for some $p < 1$, and use the characterization $1/R = \limsup |X_n|^{1/n}$.]

12. THE LAW OF THE ITERATED LOGARITHM

Throughout this section $X_1, X_2, \ldots$ will be independent random variables having a common distribution with mean 0. The strong law of large numbers (Theorem 9.4) asserts that $S_n/n \to 0$ as $n \to \infty$, or in other words that $|S_n| = o(n)$ a.s.[11] It is natural to ask how fast the convergence goes. If we assume nothing about the distribution of the X_n beyond existence of the

---

[11] Recall that $a_n = o(b_n)$ ($b_n > 0$) means that $a_n/b_n \to 0$ as $n \to \infty$, while $a_n = O(b_n)$ asserts that the ratios a_n/b_n are bounded.

first moment, then no more can be said. But it seems plausible that with some restrictions on the distribution stronger conclusions might hold, and it is remarkable that precise results about the rate of convergence can be obtained under quite general conditions.

The story begins with a special case: Suppose that X_n takes the values ± 1 with probabilities $1/2$ (the coin-tossing model). The probability space can be taken as the unit interval with Lebesgue measure, and then the nth digit b_n in the binary expansion of a number t in that interval provides a model for X_n according to the relation

$$X_n(t) = 2b_n(t) - 1.$$

[Recall Problem 8.2. The $X_n(t)$ are also known as Rademacher functions.] In this situation, statements about the partial sums S_n can be rephrased in terms of the number of 1's among the first n binary digits of t. That was the context in which the problem was first studied and where these results were obtained:

1913: $|S_n| = O(n^{\frac{1}{2}+\epsilon})$ a.s.   for any $\epsilon > 0$.   (Hausdorff)

1914: $|S_n| = O(\sqrt{n \log n})$ a.s.   (Hardy and Littlewood)

1923: $|S_n| = O(\sqrt{n \log \log n})$ a.s.   (Khintchine)

1924: $\limsup_{n \to \infty} \dfrac{|S_n|}{\sqrt{n \log \log n}} = \sqrt{2}$ a.s.   (Khintchine)

A few years later the last result was generalized by Kolmogorov to a wide class of sequences of independent random variables.

We will develop these results in their historical order, which makes the proofs seem more natural than they would appear in isolation. But there is no gain in limiting ourselves to the coin-tossing case even temporarily.

Estimate 1 (Hausdorff). *If the common distribution of the X_k has mean 0 and finite moments of all orders, then for any $\epsilon > 0$*

$$|S_n| = O(n^{\frac{1}{2}+\epsilon}) \quad \text{a.s.} \tag{1}$$

Proof. In our first pass at the law of large numbers, we showed in Eq. (8.3) (with $\mu = 0$) that $E(S_n^4) \leq Cn^2$. We then noted that

$$\sum_{n=1}^{\infty} E\left(\left[\frac{S_n}{n}\right]^4\right) \le \sum_{n=1}^{\infty} \frac{Cn^2}{n^4} < \infty,$$

from which, using Lemma 8.1, we concluded that $S_n/n \to 0$ a.s. At the same time, and with no more effort, we could have obtained a stronger result by observing that

$$\sum_{n=1}^{\infty} E\left(\left[\frac{S_n}{n^\alpha}\right]^4\right) \le \sum_{n=1}^{\infty} \frac{Cn^2}{n^{4\alpha}} < \infty \qquad (2)$$

provided that $\alpha > 3/4$. From the same lemma, we then have $S_n/n^\alpha \to 0$ for any such α.

This can be improved by using higher moments. It is not hard to verify that for any positive integer k,

$$E(S_n^{2k}) \le Cn^k, \qquad (3)$$

where C depends on k but not n. Accordingly, we can replace (2) by

$$\sum_{n=1}^{\infty} E\left(\left[\frac{S_n}{n^\alpha}\right]^{2k}\right) \le \sum_{n=1}^{\infty} \frac{Cn^k}{n^{2k\alpha}}, \qquad (4)$$

and this series converges provided $\alpha > (k+1)/2k$. The conclusion is that for any such α we again have $S_n/n^\alpha \to 0$. By choosing k large enough we can see that this result must hold for any $\alpha > 1/2$, and so $|S_n| = O(n^{\frac{1}{2}+\epsilon})$ with probability 1 for all $\epsilon > 0$. This proves Hausdorff's estimate. □

Remark. All this could just as well be done using the first part of Borel–Cantelli instead of Lemma 8.1.

Problem 1. *Prove* (3).

There are two tricks involved in the improvements to Hausdorff's estimate. Curiously, they can be introduced in either order; either one suffices for the improvement to $O(\sqrt{n \log n})$ and then adding the other makes it possible to show $O(\sqrt{n \log \log n})$. Probably the most natural way to proceed, though, is to reflect at this point that we got better and better results above by using higher and higher moments of S_n, and the "limit" to this process might be to use an exponential function. We need a replacement for the estimate (3).

Lemma 1. *Suppose that $|X_k| \le M$ holds a.s. for some constant M, and let σ^2 be the variance of the X_k. Then for any x such that $0 \le x \le 2/M$,*

$$E(e^{xS_n}) \le e^{\frac{1}{2}nx^2\sigma^2(1+xM^3/\sigma^2)}. \tag{5}$$

Proof. The random variables e^{xX_k} are independent and have a common distribution, so that

$$E(e^{xS_n}) = E(e^{xX_1})^n. \tag{6}$$

But using the boundedness of $|X_1|$ plus $E(X_1) = 0$, for $x \ge 0$ we have

$$E(e^{xX_1}) = E\left(\sum_{k=0}^{\infty} \frac{x^k X_1^k}{k!}\right)$$

$$= 1 + 0 + \frac{x^2\sigma^2}{2} + \sum_{k=3}^{\infty} \frac{x^k E(X_1^k)}{k!}$$

$$\le 1 + \frac{x^2\sigma^2}{2} + \sum_{k=3}^{\infty} \frac{x^k M^k}{k!}$$

$$\le 1 + \frac{x^2\sigma^2}{2} + \frac{x^3 M^3}{6}\left(1 + \frac{xM}{3} + \frac{x^2 M^2}{3^2} + \cdots\right)$$

$$= 1 + \frac{x^2\sigma^2}{2} + \frac{x^3 M^3}{6} \cdot \frac{1}{1-xM/3}.$$

Provided that $x \le 2/M$, the last fraction in the bottom line is no more than 3, so that the bound becomes

$$E(e^{xX_1}) \le 1 + \frac{x^2\sigma^2}{2}\left(1 + \frac{xM^3}{\sigma^2}\right) < e^{\frac{1}{2}x^2\sigma^2(1+xM^3/\sigma^2)}, \tag{7}$$

and combining (7) with (6) yields (5). □

To exploit (5) we use (6.8), the exponential form of Chebyshev's inequality.

Corollary 1. *Provided $a_n = o(n)$ as $n \to \infty$,*

$$P(S_n \ge a_n) \le \exp\left(-\frac{a_n^2}{2n\sigma^2} + O\left(\frac{a_n^3}{n^2}\right)\right). \tag{8}$$

Proof. Combining (6.8) and (5) yields

$$P(S_n \geq a_n) \leq e^{-xa_n} E(e^{xS_n}) \leq e^{-xa_n} e^{\frac{1}{2}nx^2\sigma^2(1+xM^3/\sigma^2)} \quad (9)$$

for any $x \leq 2/M$ [so that (5) holds]. Within this range x is at our disposal, and we will choose it to approximately optimize (i.e., minimize) the upper bound in (9). The choice that does this is

$$x = \frac{a_n}{n\sigma^2};$$

seeing this involves nothing more than finding the minimum of a quadratic function. [This value of x satisfies the condition $x \leq 2/M$ for large n since by assumption $a_n = o(n)$.] Substituting in (9) we obtain (8). □

Estimate 2 (Hardy and Littlewood). *If the i.i.d. random variables X_k have mean 0 and are bounded a.s. by a constant M, then*

$$|S_n| = O(\sqrt{n \log n}) \quad \text{a.s.} \quad (10)$$

Proof. Let $a_n = J\sqrt{n \log n}$, where J is any positive number; then (8) becomes

$$P(S_n \geq a_n) \leq \exp\left(-\frac{J^2 n \log n}{2n\sigma^2} + O\left(\frac{n^{3/2}(\log n)^{3/2}}{n^2}\right)\right)$$

$$= e^{-J^2 \log n/2\sigma^2 + o(1)}.$$

Hence we have

$$P(S_n \geq J\sqrt{n \log n}) \leq (1 + \epsilon) n^{-J^2/2\sigma^2}, \quad (11)$$

and the sum of the right side converges provided $J > \sigma\sqrt{2}$. Thus, for such values of J, the Borel–Cantelli lemma implies that only finitely many of the events $A_n = \{S_n \geq J\sqrt{n \log n}\}$ can occur, with probability 1. Since the whole argument can be repeated for the sums $-S_n$, the conclusion (10) is established. □

Not much more progress can be made by finding a better upper bound, since the one used above is quite accurate. The problem is that the events A_n are highly correlated, so adding up their probabilities will badly overestimate the probability of their union or lim sup no matter how accurately we estimate the terms $P(A_n)$. To get around this difficulty, we must handle the terms in large groups, so that only a thin subseries of $\sum P(A_n)$ needs to converge. This is the "second trick" mentioned above. To carry it out we need another lemma similar to Kolmogorov's inequality (Theorem 9.1).

Lemma 2. *Let $X_1, X_2, \ldots$ be independent random variables with means 0 and variances σ^2. Then, for any $a > 0$,*

$$P\left(\max_{1 \leq k \leq n} S_k \geq a\right) \leq 2 P(S_n \geq a - \sqrt{2n\sigma^2}). \tag{12}$$

Proof. We define events A and A_n almost exactly as in (9.4) and (9.5) but this time without the absolute values:

$$A = \left\{\omega : \max_{1 \leq k \leq n} S_k \geq a\right\}$$

and

$$A_k = \{\omega : S_j < a \text{ for } j = 1, \ldots, k-1 \text{ but } S_k \geq a\}.$$

Again the A_k are disjoint, and again their union is A. Now we write

$$P(A) = P(A \cap \{S_n \geq a - \sqrt{2n\sigma^2}\}) + P(A \cap \{S_n < a - \sqrt{2n\sigma^2}\})$$
$$= C + D, \tag{13}$$

say. Obviously $C \leq P(S_n \geq a - \sqrt{2n\sigma^2})$. As for the second term, we have

$$D = \sum_{k=1}^{n} P(A_k \cap \{S_n < a - \sqrt{2n\sigma^2}\}); \tag{14}$$

this partition on k is similar to one used earlier. Note next that

$$P(A_k \cap \{S_n < a - \sqrt{2n\sigma^2}\}) \leq P(A_k \cap \{S_n - S_k < -\sqrt{2n\sigma^2}\})$$
$$\leq P(A_k \cap \{|S_n - S_k| \geq \sqrt{2n\sigma^2}\});$$

the reason is that for $\omega \in A_k$ we must have $S_k \geq a$. Using independence to factor the last probability and then applying the usual form of Chebyshev's inequality (6.6), we obtain

$$P(A_k \cap \{S_n < a - \sqrt{2n\sigma^2}\}) \leq P(A_k) P(|S_n - S_k| \geq \sqrt{2n\sigma^2})$$
$$\leq P(A_k) \frac{(n-k)\sigma^2}{2n\sigma^2} \leq \frac{1}{2} P(A_k).$$

THE LAW OF THE ITERATED LOGARITHM　　　　　　　　　　　　67

Substituting in (14) yields the bound $D \leq P(A)/2$, so that (13) becomes

$$P(A) = C + D \leq P(S_n \geq a - \sqrt{2n\sigma^2}) + \frac{1}{2}P(A),$$

which is the same as (12). □

Problem 2. *Show that if the distributions of the random variables X_k in the lemma are symmetric around 0, the bound (12) can be improved to*

$$P\left(\max_{1 \leq k \leq n} S_k \geq a\right) \leq 2P(S_n \geq a). \tag{15}$$

Estimate 3 (Khintchine). *Under the same conditions as in Estimate 2,*

$$\limsup_{n \to \infty} \frac{|S_n|}{\sqrt{n \log \log n}} \leq \sqrt{2}\sigma \quad \text{a.s.} \tag{16}$$

Proof. We now define $a_n = (1 + \epsilon)\sigma\sqrt{2n \log \log n}$; then the conclusion (16) amounts to showing that only finitely many of the events $A_n = \{S_n \geq a_n\}$ occur whatever $\epsilon > 0$ we choose. (Of course we need, and will have, the same thing for the sequence $\{-S_n\}$.) It would suffice to show that $\sum P(A_n) < \infty$, but this is not true. To get around this difficulty, pick any number $d > 1$ and let $n_k = [d^k]$. (The notation $[x]$ means the largest integer not exceeding x.) Thus $\{n_k\}$ forms a roughly geometric subsequence of the integers. For this subsequence $\sum P(A_{n_k}) < \infty$ *is valid, so that*

$$P(S_{n_k} \geq a_{n_k} \text{ infinitely often}) = 0.$$

Of course this result by itself would not be good enough, since it verifies the assertion on the growth of S_n only for the subsequence $\{n_k\}$. But using this approach in combination with Lemma 2 will do the job.

From (12) we have

$$P\left(\max_{1 \leq j \leq n_k} S_j \geq a_{n_k}\right) \leq 2P(S_{n_k} \geq a_{n_k} - \sqrt{2n_k\sigma^2}). \tag{17}$$

Since a_n is of a larger order of magnitude than $\sqrt{n}$,

$$a_{n_k} - \sqrt{2n_k\sigma^2} = a_{n_k}(1 - o(1))$$

as $k \to \infty$, and so (17) can be replaced by

$$P\left(\max_{1 \le j \le n_k} S_j \ge a_{n_k}\right) \le 2P(S_{n_k} \ge a_{n_k}[1 - o(1)]). \tag{18}$$

We can estimate the right-hand side of (18) using Corollary 1 just as we did in the proof of Estimate 2; the result is

$$P\left(\max_{1 \le j \le n_k} S_j \ge a_{n_k}\right) \le 2\exp\left(-\frac{a_{n_k}^2}{2\sigma^2 n_k}[1 + o(1)]\right).$$

If we substitute the value of a_{n_k}, it is easy to check that the sum over k of the right-hand side converges, and so by the Borel–Cantelli lemma we know that only finitely many of the events A_{n_k} will occur; in other words,

$$P\left(\max_{1 \le j \le n_k} S_j < a_{n_k} \text{ for all large } k\right) = 1. \tag{19}$$

It is easy to see that (19) gives what we want. Because the function $\sqrt{x \log \log x}$ is monotonic increasing, (19) implies that, for all large k, when $n_{k-1} < j \le n_k$ we have

$$\frac{S_j}{\sqrt{j \log \log j}} < \frac{a_{n_k}}{\sqrt{n_{k-1} \log \log n_{k-1}}}.$$

But the right-hand side tends to $(1 + \epsilon)\sigma\sqrt{2d}$ as $k \to \infty$, and so for any $\eta > 0$ and all large values of j we have

$$\frac{S_j}{\sqrt{j \log \log j}} < (1 + \epsilon)\sigma\sqrt{2d}(1 + \eta)$$

with probability 1. Since each of the factors $(1 + \epsilon)$, d, and $(1 + \eta)$ can be taken as close to 1 as desired, this establishes the lim sup relation (16) with S_n in place of $|S_n|$. As before, the same proof applies to the sequence $\{-S_n\}$, so the estimate (16) holds as stated. $\square$

Problem 3. *Apply Estimates 1, 2, and 3 to the number of times a particular binary or decimal digit occurs among the first n digits of a number $x \in [0, 1]$.*

Problem 4. *Using Lemma 2 and the "second trick" plus the moment estimate (3), obtain the Hardy–Littlewood estimate without using Lemma 1.*

This approach even lets us improve on (10); *what is the best result it can yield?*

To complete the "law of the iterated logarithm," we need to show that the lim sup in (16) is not less than $\sqrt{2}\sigma$. This can be done by an argument similar to the one above once we have a lower bound to put in place of (8). One such bound is the following.

Lemma 3. *Under the conditions of Lemma 1, assume that $\{b_n\}$ are constants such that $b_n = o(n)$, but $b_n/\sqrt{n} \to \infty$. Then for any $\eta > 0$,*

$$P(S_n \geq b_n) \geq \exp\left(-\frac{b_n^2}{2n\sigma^2}(1+\eta)\right) \tag{20}$$

for all large n.

We will not prove Lemma 3, for the argument is somewhat tedious and perhaps enough is enough. [A complete proof can be found in Billingsley (1995)]. For the case of Bernoulli trials, the lemma and the whole law of the iterated logarithm are proved nicely in Feller (1968). It is worth noting that in one special case—when the $\{X_k\}$ have a normal distribution with mean 0—the bound (20) and the analogous upper bound can be verified quite easily. This is so because the sum S_n itself has a normal distribution in that case (recall Problem 3.6), again with mean 0 but with variance $\sigma^2 n$.

Problem 5. *Show that* (20) *holds for the normal case just mentioned.*

Given the bound (20), we have the following.

Estimate 4 (Khintchine). *Under the same conditions as Estimate 3, we have*

$$\limsup_{n\to\infty} \frac{S_n}{\sqrt{n \log \log n}} \geq \sqrt{2}\sigma \quad (a.s.) \tag{21}$$

Sketch of Proof. This time let $a_n = (1-\epsilon)\sigma\sqrt{2n \log \log n}$; we must show that, with probability 1, infinitely many of the events $A_n = \{S_n \geq a_n\}$ occur for any $\epsilon > 0$. To accomplish this we use the second part of Borel–Cantelli. Since this lemma needs independence, it cannot be applied to the $\{A_n\}$, not even to a subsequence of them, but we *can* apply the lemma to independent events constructed from increments of the sequence $\{S_n\}$.

Let $n_k = [D^k]$, where we have changed from d to D to suggest that this time large values will be the useful ones. Define the events

$$B_k = \{\omega : S_{n_k} - S_{n_{k-1}} \geq a_{n_k}\}.$$

These events, defined in terms of disjoint blocks of the random variables $\{X_n\}$, are independent, and using (20) we can show that if D is large enough, then

$$\sum_{k=2}^{\infty} P(B_k) = \infty. \tag{22}$$

From the Borel–Cantelli lemma, therefore, we have

$$P(S_{n_k} \geq a_{n_k} + S_{n_{k-1}} \text{ for infinitely many } k) = 1. \tag{23}$$

By Estimate 3, however, we can be sure that for any $\delta > 0$,

$$|S_{n_{k-1}}| \leq \sqrt{2}\sigma(1 + \delta)\sqrt{n_{k-1} \log \log n_{k-1}}$$

for all large k, with probability 1. Putting this fact and the values of the a_n and n_k into (23), we find that for infinitely many k,

$$S_{n_k} \geq \sqrt{2}\sigma\sqrt{n_k \log \log n_k}\left(1 - \epsilon - \frac{1+\delta}{\sqrt{D}}\right).$$

Since D can be taken arbitrarily large, this shows that (using a slightly larger value of ϵ) infinitely many A_n must occur. Therefore the lim sup in (21) is at least $\sqrt{2}\sigma$, and the theorem is proved. (In this case, omitting the absolute value makes the conclusion formally stronger.) □

Problem 6. *Verify* (22).

Remark. It was shown by Kolmogorov that (16) and (21) are valid for i.i.d. random variables assuming only that their means are 0 and the variance exists; the proof uses a truncation argument akin to the one found in Section 9 but more delicate. There are also new forms of the iterated logarithm theorem that enrich its content even in the simplest cases such as Bernoulli trials. This circle of ideas, dating to the early years of the twentieth century, continues to offer challenging problems and new insights.

CHAPTER THREE

Limiting Distributions and the Central Limit Problem

13. WEAK CONVERGENCE OF MEASURES

We have been studying some of the ways in which averages or sums of independent random variables converge to a limit. One of the great classical theorems (or cluster of theorems), the law of large numbers, was the prototype. We now turn to another collection of results typified by the famous "central limit theorem," in which the *distributions* of certain random variables tend to a limit although the variables themselves usually do not. In this preliminary section we examine the commonest and most useful notion of convergence for a sequence of probability measures. We begin with the general definition, although this chapter is mainly concerned with measures on the real line. Later on in Section 25 we will catch a glimpse of the importance of weak convergence in function spaces.

Suppose that S is a metric space and $\mathcal{S}$ is the family of all its Borel subsets; that is, $\mathcal{S}$ is the σ-field generated by the open sets of S. Let $\{\mu_n\}$ be a sequence of finite measures on $(S, \mathcal{S})$.

Definition 1. *We say that μ_n converges weakly to a finite measure μ (in symbols $\mu_n \Rightarrow \mu$), provided that*

$$\lim_{n \to \infty} \int_S f \, d\mu_n = \int_S f \, d\mu \qquad (1)$$

for every bounded, continuous function f on S.

The μ_n need not be probability measures (in our applications they always are), but choosing f to be a constant function shows that in any case $\mu_n(S) \to \mu(S)$.

Problem 1. *Let μ_n be the measure consisting of a unit mass at a point x_n [i.e., $\mu_n(E) = 1$ if $x_n \in E$, and $= 0$ otherwise]. Show that $\mu_n \Rightarrow \mu$ provided that $\lim x_n = x$ exists, and in that case μ must be the unit mass at the point x. Prove the converse too, in the case where S is the Euclidean space R^k.*

Problem 2. *Let $S = [0, 1]$, and let μ_n be the discrete measure that puts mass $1/n$ on each of the points $0, 1/n, 2/n, \ldots, n/n = 1$. Prove that $\mu_n \Rightarrow \mu$, where μ is Lebesgue measure on $[0, 1]$.*

In the case $S = R^1$, there is a simple relationship between weak convergence of measures and the corresponding distribution functions.[1]

Theorem 1. *Let P_n and P be probability measures on the Borel sets of R^1, and let F_n and F be their respective distribution functions. Then $\{P_n\}$ converges weakly to P if and only if*

$$\lim_{n \to \infty} F_n(x) = F(x) \tag{2}$$

for each x at which F is continuous.

Proof. First suppose that $P_n \Rightarrow P$. Choose any $\epsilon > 0$ and consider the bounded continuous function $g_\epsilon(t)$ that has the value 1 for $t \leq x$, 0 for $t \geq x + \epsilon$, and is linear in between. By the definition of weak convergence,

$$\lim_{n \to \infty} \int_{R^1} g_\epsilon \, dP_n = \int_{R^1} g_\epsilon \, dP. \tag{3}$$

It is obvious that

$$F_n(x) = \int_{-\infty}^{x} g_\epsilon \, dP_n \leq \int_{-\infty}^{\infty} g_\epsilon \, dP_n, \tag{4}$$

and also that

$$\int_{-\infty}^{\infty} g_\epsilon \, dP \leq F(x + \epsilon). \tag{5}$$

[1]This result is sometimes called the "Helly–Bray theorem."

WEAK CONVERGENCE OF MEASURES

Combining (3), (4), and (5), we have

$$\limsup_{n\to\infty} F_n(x) \leq F(x + \epsilon). \tag{6}$$

Letting $\epsilon \to 0$, we obtain $F(x)$ as an upper bound because of the right continuity of F. This holds for every x.

Now let $f_\epsilon(t)$ be the function that is 1 for $t \leq x - \epsilon$, 0 for $t \geq x$, and again linear in between. Arguing much as before, this time we have

$$\liminf_{n\to\infty} F_n(x) \geq \lim_{n\to\infty} \int_{R^1} f_\epsilon \, dP_n = \int_{R^1} f_\epsilon \, dP \geq F(x - \epsilon). \tag{7}$$

If we let $\epsilon \to 0$ the lower bound becomes $F(x-)$, which is the same as the upper bound $F(x)$ for the lim sup of $F_n(x)$ provided that x is a continuity point. This proves the "only if."

Corollary 1. *If a sequence of probability measures on R^1 converges weakly to P and also to P', then P and P' are identical.*

Proof. The corollary is true since P and P' must have the same distribution function by the first part of Theorem 1. (This uniqueness actually holds for weak limits in any metric space, even though this proof is tied to one dimension.)

We return to the proof of Theorem 1. To prove the "if" part, assume that (2) holds and that f is a bounded, continuous function; we must show that

$$\lim_{n\to\infty} \int_{R^1} f \, dP_n = \int_{R^1} f \, dP. \tag{8}$$

Because of (2), we can choose an interval $[-A, A]$ such that the measures P and P_n (for all n from some point on) each have weight at most ϵ outside the interval. [It is enough to choose A so that A and $-A$ are continuity points of F and so that $F(-A)$ and $1 - F(A)$ are each less than $\epsilon/3$.] Then we will have

$$\left| \int_{R^1} f \, dP_n - \int_{R^1} f \, dP \right| \leq \left| \int_{-A}^{A} f \, dP_n - \int_{-A}^{A} f \, dP \right| + 2\epsilon M \tag{9}$$

where M is an upper bound for $|f(x)|$. Since ϵ is arbitrary, this reduces the problem to showing that

$$\lim_{n\to\infty} \int_{-A}^{A} f \, dP_n = \int_{-A}^{A} f \, dP. \tag{10}$$

Equation (10), written in terms of Riemann–Stieltjes integrals, is a standard advanced-calculus result known as "Helly's second theorem." The proof is not hard. On the compact interval $[-A, A]$ the function f can be uniformly approximated by a step function; by making trifling adjustments if necessary, the points where the steps occur can be chosen to be continuity points of F. The integrals in (10) are thus approximated by finite sums, and (2) implies that the sums formed using F_n converge to those formed with F. We omit the details, which can be found in Widder (1946) or (better) that the reader can supply for herself. □

Remark 1. The proof above shows that the conclusion of weak convergence still holds if hypothesis (2) is replaced by the apparently weaker assumption that

$$\lim_{n \to \infty} [F_n(y) - F_n(x)] = F(y) - F(x) \tag{11}$$

for any finite numbers $y > x$ that are continuity points for F.

Remark 2. We will write $F_n \Rightarrow F$ to indicate that (2) holds; that is, that $F_n(x)$ converges to the distribution function $F(x)$ at all continuity points. Because of the theorem, of course, this is equivalent to the weak convergence of the corresponding measures.

One way in which weak convergence of measures on R^1 comes about is illustrated by the problem below. (Note that this is *not* the situation we will usually encounter in studying the central limit problem.)

Problem 3. *Assume that X_n and X are random variables on some probability space and that $X_n \to X$ in probability. If P_n and P are their distributions on R^1, prove that $P_n \Rightarrow P$. [Hint: Use Theorem 1.]*

The weak convergence of measures on R^1 can be described by a metric, and this will come in handy later on. Let P and Q be two probability measures on the line with corresponding distribution functions F and G. The *Lévy distance L* between F and G (or, equivalently, between P and Q) is defined by

$$L(F, G) = \inf\{h > 0 : F(x-h) - h \le G(x) \le F(x+h) + h \text{ for all } x\}. \tag{12}$$

Problem 4. *Prove that the functional L is a metric on probability measures. That is, show that it is reflexive and symmetric (easy), positive when the measures are distinct, and satisfies the triangle inequality:*

$L(F, G) \leq L(F, H) + L(H, G)$ *for any three measures having the distribution functions F, G, and H.*

Theorem 2. *Let P_n and P be probability measures on R^1. Then*

$$P_n \Rightarrow P \quad \text{if and only if} \quad L(P_n, P) \to 0. \tag{13}$$

Proof. First suppose that $L(P_n, P) \to 0$; by (12) this means that for any $h > 0$ and for all large n, we have

$$F(x - h) - h \leq F_n(x) \leq F(x + h) + h \quad \text{for all } x. \tag{14}$$

Let x be any continuity point of F. Then the outer bounds in (14) will both be close to $F(x)$ when h is small, and it follows that $F_n(x) \to F(x)$. By Theorem 1, this in turn implies that $P_n \Rightarrow P$.

To prove the converse, choose A so that $F(-A)$ and $1 - F(A)$ are both less than $h/2$. Let $\{-A = x_0, x_1, x_2, \ldots, x_N = A\}$ be a partition of the interval $[-A, A]$ such that all the intervals $[x_i, x_{i+1}]$ have lengths less than h; again we can also arrange that the partition points $\{x_i\}$ are continuity points for F. We are now assuming that $P_n \Rightarrow P$, and by Theorem 1 this means that the distribution functions converge at continuity points. For all large n, therefore, we will have

$$|F_n(x_i) - F(x_i)| \leq \frac{h}{2}, \quad i = 0, 1, \ldots, N. \tag{15}$$

For any such value of n [one for which (15) holds] it can be verified that

$$F_n(x - h) - h \leq F(x) \leq F_n(x + h) + h \quad \text{for all } x, \tag{16}$$

so that $L(F_n, F) \leq h$. This proves the theorem. □

Problem 5. *Show that* (16) *holds for all x. [Hint: It is necessary to consider separately the cases where $x \leq -A$, where $-A < x < A$, and where $x \geq A$.]*

The Lévy metric has the natural property that adding a "small" random variable to X does not move its distribution very far. A precise statement to this effect will be useful later on.

Lemma 1. *Let X be a random variable with distribution function F, and let Y be any random variable[2] with expected value 0 and standard*

---

[2]It is not assumed that Y and X are independent.

deviation σ. Suppose that F_σ denotes the distribution function of $X + Y$. Then

$$L(F, F_\sigma) \leq \sigma^{2/3}. \tag{17}$$

Proof. By Chebyshev's inequality, $P(|Y| \geq h) \leq \sigma^2/h^2$ for any $h > 0$. Hence

$$F_\sigma(x) = P(X \leq x - Y)$$
$$= P(X \leq x - Y \text{ and } |Y| \geq h) + P(X \leq x - Y \text{ and } |Y| < h)$$
$$\leq \frac{\sigma^2}{h^2} + P(X \leq x + h) = \frac{\sigma^2}{h^2} + F(x + h). \tag{18}$$

Similarly,

$$F(x) = P(X \leq x) = P(X + Y \leq x + Y)$$
$$= P(X+Y \leq x+Y \text{ and } |Y| \geq h) + P(X+Y \leq x+Y \text{ and } |Y| < h)$$
$$\leq \frac{\sigma^2}{h^2} + P(X + Y \leq x + h) = \frac{\sigma^2}{h^2} + F_\sigma(x + h). \tag{19}$$

Now combine (18) and (19) and choose $h = \sigma^{2/3}$; (17) is the result. □

Finally, it can be very useful to know when a sequence of measures has a convergent subsequence. For R^1 the answer is simple.

Theorem 3. *Let $\{F_n\}$ be the distribution functions of a sequence of probability measures on R^1. Suppose that for every $\epsilon > 0$, there exists a number A_ϵ such that*

$$F_n(-A_\epsilon) < \epsilon \quad \text{and} \quad F_n(A_\epsilon) > 1 - \epsilon \qquad \text{for all } n. \tag{20}$$

Then the sequence of measures has a weakly convergent subsequence.[3]

Proof. First we will choose a subsequence $\{F_{n'}\}$ of the distribution functions such that $F_{n'}(r)$ converges at every rational number r; to do this we use the "diagonal method" of Cantor. Arrange the rational numbers as a sequence $\{r_n\}$ and select a sequence $\{n_1\}$ of integers such that $F_{n_1}(r_1)$ converges; this is possible since $\{F_n(r_1)\}$ is a bounded sequence. Then choose a subsequence $\{n_2\}$ of $\{n_1\}$ for which $F_{n_2}(r_2)$ converges also [as

---

[3] This theorem, possibly with minor variations, is also called "Helly's theorem."

WEAK CONVERGENCE OF MEASURES 77

well as $F_{n_2}(r_1)$]. Continue this process for each rational number. Finally, let the subsequence $\{F_{n'}\}$ consist of the diagonal elements: We take the first element of $\{F_{n_1}\}$, the second element of $\{F_{n_2}\}$, and so on. This sequence is a subsequence of each of the $\{F_{n_k}\}$, and so it is clear that $F_{n'}(r)$ converges for each of the rationals $r_1, r_2, \ldots$.

For each rational number r, we now have the existence of the limit

$$L(r) = \lim_{n' \to \infty} F_{n'}(r).$$

Obviously L is nonnegative, increasing, and bounded; we define the function

$$F(x) = \inf_{r > x} L(r)$$

for all x. We will show that F is the distribution function of a probability measure that is the weak limit of the subsequence $\{P_{n'}\}$.

Clearly F is nondecreasing, its values lie between 0 and 1, and it is also easy to see that it is right-continuous. So far we have not used assumption (20). Invoking that condition, it becomes evident that $F(-A_\epsilon) < \epsilon$. Since this holds for every $\epsilon > 0$, we conclude that $F(-\infty) = 0$. Similarly we see that $F(\infty) = 1$, and so F is a distribution function. To finish the proof, it only remains to show that $F_{n'}(x) \to F(x)$ at continuity points of the latter.

Let x be a fixed point at which F is continuous. For any rational number $r > x$, we have

$$\limsup_{n' \to \infty} F_{n'}(x) \leq \lim_{n' \to \infty} F_{n'}(r) = L(r) \leq F(r).$$

If instead we take a rational $r' \leq x$, we get

$$\liminf_{n' \to \infty} F_{n'}(x) \geq \lim_{n' \to \infty} F_{n'}(r') = L(r') \geq F(r' - \epsilon)$$

for any $\epsilon > 0$. Combining these, we have

$$F(r' - \epsilon) \leq \liminf_{n \to \infty} F_{n'}(x) \leq \limsup_{n \to \infty} F_{n'}(x) \leq F(r) \qquad (21)$$

where r' is any rational $\leq x$ and r is any rational $> x$. Since F is continuous at x, this implies that $F_{n'}(x) \to F(x)$ and completes the proof. □

Remark. Under the conditions of the theorem, every infinite subsequence of $\{P_n\}$ contains a weakly convergent sub-subsequence. Such a set of measures is said to be *conditionally compact*—"conditionally" because the limiting measure need not belong to the original set.

Problem 6. *Show that the sufficient condition of Theorem 3—that is, the existence for every $\epsilon > 0$ of an A_ϵ such that (20) holds—is also necessary for conditional compactness of the sequence of measures.*

The generalization of Theorem 3 to any metric space is easily stated: A family of probability measures $\{P_n\}$ on $(S, \mathcal{S})$ is said to be *tight* if for every $\epsilon > 0$, there exists a compact set $K_\epsilon \subset S$ such that $P_n(S - K_\epsilon) \leq \epsilon$ for all n. The following result is due to Yu. V. Prohorov.

Theorem 4. *A tight sequence of probability measures is conditionally compact. When S is separable and complete, a conditionally compact sequence of measures is also necessarily tight.*

Note that for R^1 tightness reduces to (20) so that Theorem 3 is a special case of Prohorov's theorem. We will not give the proof in general; it can be found in Billingsley (1968).

14. THE MAXIMUM OF A RANDOM SAMPLE

We will now discuss briefly a class of limit theorems that are important for many applications but simple to derive. Suppose that $X_1, X_2, \ldots$ are independent random variables with the same distribution, and define

$$M_n = \max(X_1, X_2, \ldots, X_n). \tag{1}$$

It may be necessary to know the probabilities with which different values of M_n occur in the long run (when n is large). For answering this question, the nature of the distribution of the $\{X_k\}$ in the central part of its range should not be important; all that matters is the way probability is spread out in the right-hand "tail" as $x \to \infty$.

The distribution function of M_n is easily found:

$$P(M_n \leq x) = P(X_k \leq x \text{ for } k = 1, \ldots, n) = \prod_{k=1}^{n} P(X_k \leq x) = F(x)^n, \tag{2}$$

where F is the common distribution function of the X_k. Under various assumptions on F, it is then easy to find the limiting distribution for a suitable linear function of M_n. For example, if $1 - F(x) = c/x$ for large x

where $c > 0$ is a constant, then

$$P\left(\frac{M_n}{cn} \leq x\right) = F(cnx)^n = \left(1 - \frac{1}{nx}\right)^n \to e^{(-1/x)} \qquad (3)$$

for $x > 0$; the limit is 0 for $x \leq 0$. This limit function is the distribution function of a measure to which the distributions of $M_n/(cn)$ converge weakly by Theorem 13.1. More generally, we have the following.

Theorem 1. *Suppose that, for some $\alpha > 0$,*

$$\lim_{x \to \infty} x^\alpha [1 - F(x)] = c, \qquad c > 0. \qquad (4)$$

Then the distribution of $M_n/(cn)^{1/\alpha}$ converges weakly to a limit with distribution function

$$\Phi_\alpha(x) = \begin{cases} \exp(-x^{-\alpha}) & \text{for } x > 0, \\ 0 & \text{for } x \leq 0. \end{cases} \qquad (5)$$

Proof. Just as in the special case (3) above, we have for $x > 0$

$$P\left(\frac{M_n}{(cn)^{1/\alpha}} \leq x\right) = F[(cn)^{1/\alpha} x]^n = \left(1 - \frac{c + o(1)}{\{(cn)^{1/\alpha} x\}^\alpha}\right)^n \to \exp(-x^{-\alpha}),$$

and the weak convergence follows. □

Remark. Condition (4), which says that $1 - F(x)$ is asymptotic to c/x^α for large x, is sufficient but not necessary for the possibility of choosing positive constants $\{a_n\}$ such that $F(a_n x)^n$ converges weakly to (5). The result still holds for suitable a_n if the constant c is replaced by a *slowly varying function* $L(x)$; that is, a function with the property that $L(bx)/L(x) \to 1$ as $x \to \infty$ for any constant $b > 0$. [Any power of $\log(x)$ and any function tending to a positive limit are typical examples.] This turns out to be the necessary as well as sufficient condition for (5). Slowly varying functions and their close relatives *functions of regular variation*[4] were introduced by the Romanian mathematician J. Karamata in 1930 and play an important role in a number of probability limit theorems. We will not discuss them further here, however, since our survey of limiting distributions will not include the search for necessary conditions. A nice exposition and many applications of these functions can be found in Feller (1966).

---

[4]Functions of regular variation $R(x)$ are defined by the property that $\lim_{x \to \infty} R(bx)/R(x)$ exists for every $b > 0$; this implies that $R(x) = x^\alpha L(x)$ for some α, where L is slowly varying.

Different assumptions on the common distribution function F lead to different limiting distributions. When the random variables X_n are bounded above, M_n almost surely approaches the least such bound, that is, $\min\{t : F(t) = 1\}$, as n becomes large. If the differences between M_n and this limit are suitably magnified, a limiting distribution will sometimes exist.

Theorem 2. *Suppose that $F(x_0) = 1$ and that for some $\alpha > 0$,*

$$\lim_{x \to x_0^-} (x_0 - x)^{-\alpha}[1 - F(x)] = c, \qquad c > 0. \tag{6}$$

Then the distribution of $(cn)^{1/\alpha}(M_n - x_0)$ converges weakly to a limit with distribution function

$$\Psi_\alpha(x) = \begin{cases} \exp(-|x|^\alpha) & \text{for } x < 0, \\ 1 & \text{for } x \geq 0. \end{cases} \tag{7}$$

The proof is similar to that of Theorem 1 and we will leave it for the reader.

Problem 1. *Supply the proof of Theorem 2.*

In the examples considered so far, we had to multiply M_n or $(M_n - x_0)$ by a rescaling factor to obtain a limiting distribution. If the probability in the tail of $F(x)$ tends to 0 exponentially fast, however, translation of M_n, possibly plus scaling, will be necessary. For example, if the $\{X_k\}$ have a distribution such that

$$\lim_{x \to \infty} e^x [1 - F(x)] = c, \qquad c > 0, \tag{8}$$

then it is easy to see that

$$\lim_{n \to \infty} P[M_n - \log(cn) \leq x] = \exp(-e^{-x}) \qquad \text{for all } x. \tag{9}$$

The limiting distribution function in (9) is traditionally denoted $\Lambda(x)$.

Problem 2. *Verify (9).*

Since the "normal" distribution is of special interest (see the following two sections of this chapter), we include it in the form of another problem.

Problem 3. *Suppose that the distribution function of the $\{X_k\}$ is the standard normal function $N(x)$, and choose*

$$a_n = \frac{1}{\sqrt{2 \log n}}, \qquad b_n = \sqrt{2 \log n} - \frac{\log(4\pi \log n)}{2\sqrt{2 \log n}}. \qquad (10)$$

Show that in this case we again have

$$\lim_{n \to \infty} P\left(\frac{M_n - b_n}{a_n} \leq x\right) = \exp(-e^{-x}) \qquad \text{for all } x. \qquad (11)$$

Hint: An adequate approximation for the probability in the tail of the normal distribution (2.18) can be found by integrating by parts; the result is

$$1 - N(x) = \frac{1 + o(1)}{\sqrt{2\pi} x} e^{-x^2/2} \qquad \text{as } x \to \infty. \qquad (12)$$

These examples suggest a rather chaotic situation in which nearly anything can happen. This is not the case, however; it turns out that the three forms of limiting distribution we have discovered, namely Φ_α, Ψ_α, and Λ, are the only ones possible. We will return to these limiting distributions and prove the result just stated in Section 18.

Remarks. The three classes of limiting distributions for maxima were discovered during the 1920s by M. Fréchet, R. A. Fisher, and L. H. C. Tippett. In 1943 B. Gnedenko gave a systematic exposition of limiting distributions for the maximum of a random sample.[5] In that paper Gnedenko completed the proof, outlined (with unnecessary assumptions) by Fisher and Tippett, that no other limiting distributions are possible; he also investigated the conditions on F under which a particular limit appears and derived laws of large numbers for the maximum. All of this plus many more recent results can be found in the book by J. Galambos (1987).

15. CHARACTERISTIC FUNCTIONS

The "central limit problem" is the quest to determine what happens to the distribution of a sum $S_n = X_1 + \cdots + X_n$ of independent random variables,

---

[5]"Sur la distribution limité du terme maximum d'une série aléatoire," *Annals Math.* 44 (1943), pp. 423–453.

suitably standardized by a linear change of scale, when the number of terms in the sum tends to infinity. In the case of Bernoulli trials, the answer was given by the famous limit theorem of de Moivre, a result nearly as old as the first law of large numbers. One form of the theorem can be stated as follows: Let $X_1, X_2, \ldots$ be independent random variables each taking the values 1 or 0 with probabilities p and $q = 1 - p$, respectively ($p \neq 0$ or 1). Then as $n \to \infty$, the distribution of $(S_n - np)/\sqrt{npq}$ converges weakly to the "standard normal" distribution whose distribution function[6] is

$$N(x) = \frac{1}{\sqrt{2\pi}} \int_{-\infty}^{x} e^{-u^2/2} du \qquad (1)$$

By virtue of Theorem 13.1, this weak convergence is equivalent to

$$\lim_{n \to \infty} P\left(\frac{S_n - np}{\sqrt{npq}} \leq x\right) = N(x) \qquad \text{for all } x; \qquad (2)$$

(2) is the usual way of stating the conclusion. This result is the earliest example of the *central limit theorem*.

In principle, the distribution of the sum S_n can always be obtained from the distributions of the individual X_k by repeated convolution (Theorem 3.2), an operation that is usually difficult or impossible to carry out directly. In a few cases it *can* be done explicitly for all n, and then a direct evaluation of a limit like that in (2) becomes feasible. For Bernoulli trials, of course, S_n has the binomial distribution (6.2) and a careful analysis of the terms with the aid of Stirling's formula for $n!$ yields a proof of (2); this is done in Feller (1968). Other distributions for the X_k such that the distribution of S_n can be found explicitly include the Poisson, the exponential and gamma family, and the normal distribution itself. (See Problem 3.6.)

To get general results, however, some other approach is necessary. One method is to calculate the moments of S_n (as we did in Section 12) and use them to study the limit of the distribution of $(S_n - a_n)/b_n$; this idea was exploited by Chebyshev in the late nineteenth century. We will use instead a method based on *characteristic functions,* a kind of Fourier transform. This approach was pioneered by Lyapunov around 1900 and has been widely used ever since.

Definition 1. *Let X be any random variable. The function*

$$\phi(\lambda) = E(e^{i\lambda X}), \qquad (3)$$

---

[6]Normal distributions have already appeared in Sections 2, 3, 4, 12, and 14.

which is defined (at least) for all real λ, is called the characteristic function of X. (We will sometimes write ϕ_X instead of ϕ to show explicitly the random variable in question.)

If the random variable has the distribution P with distribution function F, we can rewrite the above definition as

$$\phi(\lambda) = \int_{R^1} e^{i\lambda x} \, dP(x) = \int_{-\infty}^{\infty} e^{i\lambda x} \, dF(x); \qquad (4)$$

the last integral can be understood in the Riemann–Stieltjes sense since the integrand is continuous.

Problem 1. *Show that $\phi(0) = 1$, that $|\phi(\lambda)| \leq 1$ for all real λ, and that ϕ is uniformly continuous on the real line.*

If $X_1, \ldots, X_n$ are independent, so are the random variables $e^{i\lambda X_k}$. From Theorem 3.1 and its corollary, we obtain the crucial relation

$$\phi_{S_n}(\lambda) = E(e^{i\lambda S_n}) = \prod_{k=1}^{n} E(e^{i\lambda X_k}) = \prod_{k=1}^{n} \phi_{X_k}(\lambda). \qquad (5)$$

It is this fact that makes characteristic functions so valuable for attacking the central limit problem, since it provides a tool for studying the distribution of S_n when that of each X_k is known. In the language of analysis, the characteristic function is just the Fourier (Stieltjes) transform of the distribution function F, and (5) expresses the well-known fact that the transform of a convolution of distributions is the product of the corresponding transforms.

To preview how characteristic functions are used, we will outline a proof by this method of the de Moivre limit theorem. First, we need the characteristic function of the normal distribution; for later use, we take the one with mean 0 but general variance σ^2. The result is

$$\phi(\lambda) = \frac{1}{\sqrt{2\pi}\sigma} \int_{-\infty}^{\infty} e^{i\lambda x} e^{-x^2/2\sigma^2} \, dx = e^{-\sigma^2 \lambda^2/2}. \qquad (6)$$

Problem 2. *Derive (6). [Hint: Complete the square in the exponent and use the known fact that the integral of a normal density is 1. Note that justifying this calculation requires some appeal to complex function theory.]*

Next, we find the characteristic function of $(S_n - np)/\sqrt{npq}$. By the definition of the Bernoulli random variables and (5), we have

$$\phi_{X_k}(\lambda) = pe^{i\lambda} + q, \qquad \phi_{S_n}(\lambda) = (pe^{i\lambda} + q)^n.$$

Hence,

$$\phi_n(\lambda) = E(e^{i\lambda(S_n - np)/\sqrt{npq}}) = e^{-i\lambda np/\sqrt{npq}} (pe^{i\lambda/\sqrt{npq}} + q)^n.$$

It is not hard to pass to the limit; the result is

$$\lim_{n\to\infty} \phi_n(\lambda) = e^{-\lambda^2/2} \qquad \text{for each } \lambda. \tag{7}$$

The right-hand side is, by (6), the characteristic function of the standard normal distribution $N(x)$. The proof of the theorem, then, will be accomplished if we can show that the convergence of the characteristic functions (7) implies the weak convergence of the corresponding distributions. This is called the *continuity theorem* and will be proved below. When the ground has been thus prepared, much more general results will follow with very little extra labor.

Problem 3. *Verify (7). [Hint: Use the Taylor series.]*

As a first step, we need to show that the characteristic function of a probability measure deserves its name—that is, that it uniquely determines (characterizes) the measure. That follows from this *inversion formula*.

Theorem 1. *Let ϕ and F, respectively, be the characteristic and distribution functions of a probability measure on R^1. Let $\alpha < \beta$ be two points at which F is continuous. Then*

$$F(\beta) - F(\alpha) = \lim_{\sigma \to 0+} \frac{1}{2\pi} \int_{-\infty}^{\infty} \phi(\lambda) e^{-\sigma^2 \lambda^2/2} \frac{e^{-i\lambda\beta} - e^{-i\lambda\alpha}}{-i\lambda} d\lambda. \tag{8}$$

Proof. Where does formula (8) come from? If the probability measure in question has a density $f(x)$, we have

$$\phi(\lambda) = \int_{-\infty}^{\infty} e^{i\lambda u} f(u)\, du,$$

which is essentially the ordinary Fourier transform of f. If we assume in addition that ϕ is integrable, the function f can be recovered by the Fourier inversion formula as follows:

$$f(x) = \frac{1}{2\pi} \int_{-\infty}^{\infty} \phi(\lambda) \, e^{-i\lambda x} \, d\lambda. \tag{9}$$

But this cannot be taken literally in general since there may not be a density, and even if there is one, the function ϕ may not be integrable. How can we make sense of this approach?

There is a neat way around these difficulties: We convolve the distribution function F with a normal distribution N_σ having mean 0 and variance σ^2. When σ^2 is very small this doesn't change the distribution much, but it introduces a convergence factor that makes the analysis go without difficulty. We therefore write $F_\sigma = F \star N_\sigma$ for the convolution in question; according to Theorem 3.2 this convolution represents the distribution of the sum of two independent random variables with distributions F and N_σ. Since the normal distribution is absolutely continuous, Corollary 3.2 tells us that the convolution F_σ is also absolutely continuous with density given by

$$F'_\sigma(x) = f_\sigma(x) = \int_{-\infty}^{\infty} \frac{1}{\sqrt{2\pi}\sigma} e^{-(x-u)^2/2\sigma^2} \, dF(u). \tag{10}$$

The characteristic function corresponding to (10) is the product

$$\phi_\sigma(\lambda) = e^{-\sigma^2\lambda^2/2} \phi(\lambda). \tag{11}$$

The modified characteristic function (11) is obviously integrable, so we can take its Fourier transform as in the right-hand side of (9):

$$\frac{1}{2\pi} \int_{-\infty}^{\infty} e^{-i\lambda x} \phi_\sigma(\lambda) \, d\lambda = \frac{1}{2\pi} \int_{-\infty}^{\infty} e^{-i\lambda x} e^{-\sigma^2\lambda^2/2} \int_{-\infty}^{\infty} e^{i\lambda u} \, dF(u) \, d\lambda. \tag{12}$$

Fubini's theorem applies to the iterated integral in (12), which can be rewritten in the form

$$\int_{-\infty}^{\infty} \frac{1}{2\pi} \int_{-\infty}^{\infty} e^{i\lambda(u-x)} e^{-\sigma^2\lambda^2/2} \, d\lambda \, dF(u) = \int_{-\infty}^{\infty} A(u, x) \, dF(u).$$

The inner integral here—the factor $A(u, x)$—can be evaluated easily, for it is just the characteristic function of a normal density with variance σ^{-2}, evaluated at the point $u - x$ and multiplied by the constant $(2\pi\sigma^2)^{-1/2}$. From (6) we know what that characteristic function is, so we have

$$A(u, x) = \frac{1}{\sqrt{2\pi}\sigma} e^{-(x-u)^2/2}.$$

Substituting above, we get

$$\frac{1}{2\pi}\int_{-\infty}^{\infty} e^{-i\lambda x}\phi_\sigma(\lambda)\,d\lambda = \int_{-\infty}^{\infty}\frac{1}{\sqrt{2\pi}\sigma}e^{-(x-u)^2/2\sigma^2}\,dF(u). \tag{13}$$

But this last expression is the density f_σ that we saw in (10). This proves that in the case of F_σ, that is, of any distribution F convolved with a normal distribution, the classical Fourier inversion formula (9) gives the correct expression for its density.

Since the right-hand side of (13) is the derivative of F_σ, we have

$$F_\sigma(\beta) - F_\sigma(\alpha) = \int_\alpha^\beta f_\sigma(x)\,dx = \int_\alpha^\beta \frac{1}{2\pi}\int_{-\infty}^{\infty} e^{-i\lambda x}\phi_\sigma(\lambda)\,d\lambda\,dx.$$

Again Fubini's theorem applies, so that the iterated integral can be reversed and simplified:

$$F_\sigma(\beta) - F_\sigma(\alpha) = \frac{1}{2\pi}\int_{-\infty}^{\infty}\phi(\lambda)e^{-\sigma^2\lambda^2/2}\frac{e^{-i\lambda\beta}-e^{-i\lambda\alpha}}{-i\lambda}\,d\lambda. \tag{14}$$

Therefore (8) and the theorem are established as soon as we verify that

$$\lim_{\sigma\to 0+} F_\sigma(x) = F(x) \tag{15}$$

at each x for which F is continuous. But (15) follows immediately from Lemma 13.1 which states that the Lévy distance between F and F_σ is at most $\sigma^{2/3}$ for any distribution F, plus Theorem 13.2. (With a little more work, (15) can also be proved directly.) □

The claimed uniqueness is now clear, since a probability measure is determined by the values of its distribution function at continuity points [or by the differences of its values on intervals (α, β).] Hence we have the following.

Corollary 1. *Two probability measures on the Borel sets of R^1 that have the same characteristic function are identical.*

Corollary 2. *A characteristic function is real if and only if the corresponding distribution is symmetric about zero.*

Proof. Suppose X is a random variable with characteristic function ϕ; then $-X$ has as its characteristic function the complex conjugate of ϕ. (Verify.) Because of the uniqueness, ϕ is equal to its conjugate (and hence real) if and only if X and $-X$ have the same distribution. □

Corollary 3. *If $\phi(\lambda)$ is the characteristic function of a probability measure P and if ϕ is integrable on R^1, then P is absolutely continuous and its density is given by the integral*

$$f(x) = \frac{1}{2\pi} \int_{-\infty}^{\infty} e^{-i\lambda x} \phi(\lambda) \, d\lambda. \tag{16}$$

Proof. Since $\phi \in L_1$, we can use the dominated convergence theorem to take the limit in (8) under the integral sign; the result is

$$F(\beta) - F(\alpha) = \frac{1}{2\pi} \int_{-\infty}^{\infty} \phi(\lambda) \frac{e^{-i\lambda\beta} - e^{-i\lambda\alpha}}{-i\lambda} \, d\lambda. \tag{17}$$

But the function $f(x)$ defined in (16) can be integrated from α to β and by Fubini's theorem the order of the two integrations can be switched. This yields the right side of (17); in other words,

$$\int_{\alpha}^{\beta} f(x) \, dx = F(\beta) - F(\alpha). \tag{18}$$

Thus f is the density of P; since f is continuous, it also follows from (18) that $f(x) = F'(x)$ everywhere. $\square$

Remark. If ϕ is integrable, the function f in (16) is uniformly continuous on R^1. There are, therefore, many absolutely continuous probability distributions with nonintegrable characteristic functions—to find examples, we simply construct densities that are not uniformly continuous. (Try out the case of a uniform density, for example.)

Problem 4. *Use characteristic functions to give another proof that the sum of independent, normally distributed random variables is itself normal (Problem 3.6).*

Problem 5 (The Cauchy Distribution). *Show that the function*

$$f(x) = \frac{1}{\pi} \frac{c}{c^2 + x^2}, \qquad c > 0, \tag{19}$$

is the density of a probability distribution with characteristic function

$$\phi(\lambda) = e^{-c|\lambda|}. \tag{20}$$

If $X_1, \ldots, X_n$ are independent random variables with this distribution, what is the distribution of S_n/n? [Note that the law of large numbers does not hold.]

Problem 6. *In the x-y plane, suppose that a half-line or ray originates from the origin, and that the angle θ between the ray and the x-axis is random with a uniform distribution on $[0, \pi]$. Let X be the x-coordinate of the point where the ray intersects the line $y = c$, $c > 0$. Show that the distribution of the random variable X is Cauchy with parameter c; that is, it has the density (19).*

The next task is to study the continuity of the one-to-one correspondence between distributions and characteristic functions. It is obvious from the definition of weak convergence that if $P_n \Rightarrow P$, then $\phi_n(\lambda) \to \phi(\lambda)$ for each λ; the harder and more useful step is to develop a converse.

Theorem 2. *Let F_n, F be the distribution functions of probability measures on R^1 and let ϕ_n, ϕ be the corresponding characteristic functions. Suppose that $\phi_n(\lambda) \to \phi(\lambda)$ for each fixed λ. Then $F_n \Rightarrow F$.*

Proof. Using the trick from the proof of Theorem 1, let $F_{n,\sigma}$ and F_σ be the distribution functions F_n and F (respectively) convolved with a normal distribution that has mean 0 and variance σ^2. Then by (14) above, we have

$$F_{n,\sigma}(\beta) - F_{n,\sigma}(\alpha) = \frac{1}{2\pi} \int_{-\infty}^{\infty} \phi_n(\lambda) e^{-\sigma^2 \lambda^2 / 2} \frac{e^{-i\lambda\beta} - e^{-i\lambda\alpha}}{-i\lambda} d\lambda,$$

as well as the analogous formula relating F_σ and ϕ. For a fixed value of $\sigma > 0$ the dominated convergence theorem applies, so that

$$\lim_{n \to \infty} [F_{n,\sigma}(\beta) - F_{n,\sigma}(\alpha)] = F_\sigma(\beta) - F_\sigma(\alpha).$$

By Theorem 13.1 this implies the weak convergence of the distributions $F_{n,\sigma}$ to F_σ for any $\sigma > 0$.

The fact that F_n converges to F is now easily verified using the Lévy distance. By the triangle inequality, we have

$$L(F_n, F) \leq L(F_n, F_{n,\sigma}) + L(F_{n,\sigma}, F_\sigma) + L(F_\sigma, F).$$

For any $\sigma > 0$, the central term tends to 0 as n increases because of the weak convergence of $F_{n,\sigma}$ to F_σ. But by Lemma 13.1 the first and the third terms are each bounded by $\sigma^{2/3}$. Choosing σ small, we therefore can ensure that the distance from F_n to F is arbitrarily small for all large enough n, and that finishes the proof. □

CHARACTERISTIC FUNCTIONS

Problem 7. *Using Theorem 2, show that $(S_n - n/2)/\sqrt{n}$ has a limiting normal distribution when $S_n = X_1 + \cdots + X_n$ and the X_k are independent random variables each having a uniform distribution on $[0, 1]$. (That is, X_k has a density which is 1 for x in the unit interval and 0 otherwise.)*

The continuity theorem we have derived is adequate for many applications, but an improvement on it will sometimes be needed. Suppose we know that the characteristic functions ϕ_n converge to a limit, but are not told that this limit is the characteristic function of some distribution. What can be concluded then?

Theorem 3. *Let P_n be probability distributions on R^1 with characteristic functions ϕ_n. Suppose that*

$$\lim_{n \to \infty} \phi_n(\lambda) = \phi(\lambda) \qquad \text{exists for all real } \lambda, \tag{21}$$

and that the limit function ϕ is continuous at the point $\lambda = 0$. Then ϕ is the characteristic function of a probability measure P, and $P_n \Rightarrow P$.

Proof. The continuity of $\phi(\lambda)$ implies that the measures $\{P_n\}$ form a tight family; that is, they satisfy (13.20). To see this, we will first prove the following.

Lemma 1. *Let X be a random variable with characteristic function ψ. Then for any $u > 0$,*

$$P\left(|X| > \frac{2}{u}\right) \leq \frac{1}{u} \int_{-u}^{u} [1 - \psi(\lambda)] \, d\lambda. \tag{22}$$

Proof. To establish (22), note first that the right-hand side can be written in the form

$$\frac{1}{u} \int_{-u}^{u} \int_{-\infty}^{\infty} (1 - e^{i\lambda x}) \, dF(x) \, d\lambda$$

where F is the distribution function of X. If we interchange the order of integration, this expression simplifies to

$$2 \int_{-\infty}^{\infty} \left(1 - \frac{\sin ux}{ux}\right) dF(x)$$

(and so it must be real). Since the integrand is never negative, we can obtain the estimate we seek as follows:

$$\frac{1}{u}\int_{-u}^{u}[1-\psi(\lambda)]\,d\lambda \geq 2\int_{-\infty}^{-2/u}+2\int_{2/u}^{\infty}\left(1-\frac{\sin ux}{ux}\right)dF(x)$$

$$\geq 2\int_{-\infty}^{-2/u}+2\int_{2/u}^{\infty}\left(1-\frac{1}{|ux|}\right)dF(x) \geq F\left(-\frac{2}{u}\right)+\left[1-F\left(\frac{2}{u}\right)\right],$$

which proves the lemma.

Returning to the proof of Theorem 3, since the limiting function ϕ in (21) is continuous at 0 and $\phi(0) = 1$, for any $\epsilon > 0$ we can choose u so that

$$0 \leq \frac{1}{u}\int_{-u}^{u}[1-\phi(\lambda)]\,d\lambda < \frac{\epsilon}{2}.$$

Then, using the bounded convergence theorem, we have for all large n

$$\frac{1}{u}\int_{-u}^{u}[1-\phi_n(\lambda)]\,d\lambda < \epsilon. \tag{23}$$

Combining (23) and (22), we see that the sequence of distributions $\{P_n\}$ is tight.

It is easy to complete the proof. By Theorem 13.3, there must be a subsequence of the P_n that converges weakly to some measure P. For this subsequence, the characteristic functions converge to the characteristic function of P, which therefore coincides with the limit function ϕ. We can now invoke Theorem 2 to conclude that the entire sequence $\{P_n\}$ converges to P. □

Remark. Once we had shown that the sequence of measures is tight, the proof of Theorem 3 could have been finished without appealing to Theorem 2 by using (only) the uniqueness result. In the previous paragraph, if $\{P_n\}$ does not converge to P, then by definition there is some bounded continuous function f for which

$$\int_{R^1} f\,dP_n \not\to \int_{R^1} f\,dP.$$

We can choose a subsequence of measures for which $\int f\,dP_n$ has a limit L that is *not* $\int f\,dP$. From that subsequence, choose a subsubsequence

$\{P_{n'}\}$ of measures that converge weakly to some limit. This limit must be different from P because it gives a different result when integrating the function f; nevertheless, the limiting measure has the same characteristic function ϕ as has P. This situation contradicts the uniqueness theorem (Corollary 1 above) for characteristic functions.

Postscript 1: Laplace Transforms

The theory of characteristic functions applies to all random variables. Nevertheless, it is not always the most convenient tool to use. If the random variable X is nonnegative a.s., the Laplace–Stieltjes transform or *moment-generating function* defined by

$$\theta(s) = E(e^{-sX}) = \int_0^\infty e^{-sw} dP(w) \tag{24}$$

offers a number of advantages and can be used to establish weak convergence in much the same fashion as the characteristic function ϕ. We will outline the proof.

When $X \geq 0$ a.s. (or what is the same thing, when P is supported on the nonnegative reals), it is not hard to see that (24) defines a function that is analytic for $\Re(s) > 0$ and continuous for $\Re(s) \geq 0$. If the values of such a function are known on the positive real axis, or even for any open interval of that axis, the function is uniquely determined throughout the half-plane by the principle of analytic continuation. But along the imaginary axis, the Laplace transform at $s = -i\lambda$ coincides with the characteristic function at λ—and these values, as we know from Theorem 1, uniquely determine the measure P. Thus we have proved the following:

Theorem 4. *Two probability measures on R^+ whose moment-generating functions (Laplace transforms) coincide on the positive reals are identical.*

It is almost as easy to adapt the continuity theorem to this context. Here is the analog of Theorem 3.

Theorem 5. *Let $\{P_n\}$ be probability distributions on R^+ with Laplace transforms θ_n. Suppose that*

$$\lim_{n\to\infty} \theta_n(s) = \theta(s) \quad \text{exists for all (real) } s \geq 0, \tag{25}$$

and that the limit function $\theta(s)$ is right-continuous at the point $s = 0$. Then θ is the Laplace transform of a probability measure P, and $P_n \Rightarrow P$.

Sketch of Proof. Once again, the continuity of θ at 0 implies that the measures $\{P_n\}$ form a tight family; this can be seen from an estimate similar to (22) above:

Lemma 2. *The sequence of measures $\{P_n\}$ is tight.*

Problem 8. *Prove this lemma. [Hint: Try an approach similar to the first part of the proof of Theorem 3, including an adaptation of Lemma 1.]*

Once the tightness of the $\{P_n\}$ is established, the alternative ending to the proof of Theorem 3, which was given as a remark above, applies without change and completes the proof here as well. □

Finally, as an application and exercise, the reader can prove the following version of Poisson's limit theorem for trials with unequal probabilities.

Theorem 6. *For each n, let $X_1^{(n)}, \ldots, X_{k(n)}^{(n)}$ be independent random variables taking the values 0 or 1, with $P(X_k^{(n)} = 1) = p_k^{(n)}$ for $k = 1, 2, \ldots, k(n)$. Suppose that*

$$\lim_{n \to \infty} \sum_{k=1}^{k(n)} p_k^{(n)} = \mu, \quad 0 < \mu < \infty, \quad \text{and that} \quad \lim_{n \to \infty} \max_k p_k^{(n)} = 0. \quad (26)$$

Then as $n \to \infty$, the distribution of $S_n = X_1^{(n)} + \cdots + X_{k(n)}^{(n)}$ converges to the Poisson distribution with parameter μ:

$$\lim_{n \to \infty} P(S_n = k) = e^{-\mu} \frac{\mu^k}{k!}, \quad k = 0, 1, 2, \ldots. \quad (27)$$

Problem 9. *Use Laplace transforms to prove Theorem 6. (This can also be done with characteristic functions, but the proof by Laplace transforms is neater.) [Hint: Of course, it is necessary to compute the transform of a Poisson distribution. After that, it may be helpful to first prove the theorem in the familiar special case where $k(n) = n$ and all the $p_k^{(n)}$ are equal to μ/n.]*

Postscript 2: Higher Dimensions

Characteristic functions are also valuable for studying probability distributions on spaces of several dimensions and, indeed, on more general locally compact topological groups. Their definitions and properties in finite-dimensional spaces are similar to those in R^1. In this book, we use nothing but the uniqueness theorem in the case of R^k (needed in Section 22), and this result is sketched below. The continuity theorem and other results are also valid and can serve in developing a theory of sums of independent random vectors.

Definition 2. *Let X be a random column vector in R^k with distribution P, and λ a column vector of real constants. Then the Fourier transform*

$$\phi(\lambda) = E(e^{i\lambda^T X}) = \int_{R^k} e^{i\lambda^T \mathbf{x}} \, dP(\mathbf{x}) \qquad (28)$$

is called the characteristic function *of X or of P.*

[Here, λ^T means the transpose of λ (and hence a row vector), and $\mathbf{x}$ denotes a column vector with components $x_1, \ldots, x_k$.] The essential conclusion of Theorem 1, namely uniqueness, extends easily to this setting.

Theorem 7. *The characteristic function $\phi(\lambda)$ uniquely determines P.*

Sketch of Proof. We imitate closely the proof of Theorem 1. Define a normal distribution on R^k (not the most general one) by letting each coordinate be independently normally distributed with mean 0 and variance σ^2. Once again, we let P_σ be the measure obtained by convolving P with this normal measure; the operation of convolution, just as in one dimension, corresponds to the addition of independent random vectors. This has again the effect of multiplying the corresponding characteristic functions, so we find that the characteristic function of P_σ is given by

$$\phi_\sigma(\lambda) = e^{-\frac{1}{2}\sigma^2 \lambda^T \lambda} \, \phi(\lambda).$$

Following the proof of Theorem 1, we find that P_σ has a density f_σ given by

$$f_\sigma(\mathbf{x}) = \frac{1}{(2\pi)^k} \int_{R^k} e^{-i\lambda^T \mathbf{x}} \, \phi_\sigma(\lambda) \, d\lambda.$$

(Here, $d\lambda$ means Lebesgue measure in R^k.) Thus, P_σ is uniquely determined for each $\sigma > 0$ by the characteristic function ϕ.

It only remains to show that P is determined by the family $\{P_\sigma\}$. It is again true, as in one dimension, that there is a metric on the space of probability measures such that convergence in this metric is equivalent to weak convergence. However, it is not necessary to develop that in order to draw the needed conclusion. If g is a continuous function with compact support, it is not hard to show directly that

$$\lim_{\sigma \to 0} \int_{R^k} g \, dP_\sigma = \int_{R^k} g \, dP,$$

so that the integral of g is also determined by ϕ. It follows easily that P is itself determined; this completes the proof of Theorem 7. □

16. THE CENTRAL LIMIT THEOREM

With the tools from the previous sections in hand, it is easy to prove a far-reaching generalization of the de Moivre limit theorem. We need one more preliminary fact.

Lemma 1. *Let X be any random variable with characteristic function ϕ. If the moment $E(|X^k|)$ exists for some positive integer k, then ϕ has a continuous kth derivative and*

$$\phi^{(k)}(0) = i^k E(X^k). \tag{1}$$

Proof. Let $k = 1$. Writing ϕ in the form (15.4) we have

$$\phi'(\lambda) = \lim_{h \to 0} \int_{-\infty}^{\infty} \frac{e^{ihx} - 1}{h} e^{i\lambda x} \, dP(x) \tag{2}$$

provided the limit exists. The fraction in the integrand approaches the limit ix as $h \to 0$. Moreover, it is not hard to verify that for all real λ the integrand is bounded in absolute value by $|x|$; since this function is by assumption integrable, the dominated convergence theorem applies and so integral and limit can be interchanged in (2). The result is

$$\phi'(\lambda) = i \int_{-\infty}^{\infty} e^{i\lambda x} x \, dP(x), \tag{3}$$

which is easily seen to be continuous. Setting $\lambda = 0$ yields (1) for $k = 1$. If higher moments exist, the process can be repeated to obtain the kth derivative. □

Remark. We will not prove it now, but it is sometimes useful to know that the converse assertion—that the existence of $\phi^{(k)}(0)$ implies that of $E(|X^k|)$—is true for even values of k but fails for odd ones. (See Problems 17.5 and 17.6.)

THE CENTRAL LIMIT THEOREM

Theorem 1 (Central Limit Theorem). *Suppose that $X_1, X_2, \ldots$ are independent random variables with a common distribution having mean μ and (finite) variance σ^2, and let $S_n = X_1 + \cdots + X_n$ denote their partial sum. Then*

$$F_n(x) = P\left(\frac{S_n - n\mu}{\sigma\sqrt{n}} \leq x\right) \Rightarrow N(x). \tag{4}$$

Proof. We can assume $\mu = 0$, since in any case that is true of the random variables $X_k - \mu$ whose partial sums are $S_n - n\mu$. Then by the multiplicative property (15.5), the characteristic function that corresponds to the left side of (4) is

$$\int_{-\infty}^{\infty} e^{i\lambda x} dF_n(x) = \phi_n(\lambda) = E(e^{i\lambda S_n/\sigma\sqrt{n}}) = \phi\left(\frac{\lambda}{\sigma\sqrt{n}}\right)^n, \tag{5}$$

where ϕ is the common c.f. of the X_k. But using Lemma 1 and the assumption $\mu = 0$, we have $\phi'(0) = 0$, $\phi''(0) = -\sigma^2$ and $\phi''(\lambda)$ is continuous, so that ϕ has the Taylor expansion

$$\phi(\lambda) = 1 - \frac{\sigma^2 \lambda^2}{2} + o(\lambda^2)$$

near the origin. Combining this with (5) yields

$$\phi_n(\lambda) = \left(1 - \frac{\lambda^2}{2n} + o\left(\frac{1}{n}\right)\right)^n,$$

from which it is evident that

$$\lim_{n \to \infty} \phi_n(\lambda) = e^{-\lambda^2/2} \tag{6}$$

for every fixed λ. The limit in (6) is the characteristic function of the standard normal distribution $N(x)$ by (15.6), and so according to the "continuity theorem" (Theorem 15.2) the distributions corresponding to ϕ_n must converge weakly to the normal. This completes the proof. The limit theorem discovered by de Moivre almost 300 years ago has come a long way. □

Having generalized de Moivre's central limit theorem so far, what's left to do? The theory developed up to this point can be generalized in many ways. One of these is to replace the real-valued random variables $\{X_k\}$ with variables taking values in a (usually commutative) topological group; the harmonic analysis of the group then plays a role analogous to that of the characteristic function. There is a big difference between compact and noncompact groups. We will not go into this in general, but will just consider one interesting and suggestive example, the unit circle in the complex plane.

Let $C = \{e^{i\theta}\}$ denote the complex numbers with absolute value 1. The group operation, naturally, is multiplication. Suppose that $X_1, X_2, \ldots$ are independent, identically distributed random variables with values in C; they can be represented as $X_k = e^{2\pi i Y_k}$, where the Y_k are i.i.d. variables on the interval $[0, 1)$. We will consider

$$Z_n = \prod_{k=1}^{n} X_k, \tag{7}$$

which, of course, also represents a random element of C; this product can be thought of as the sum $\sum_{k=1}^{n} 2\pi Y_k$ "wrapped around" the circle (i.e., modulo 2π). Our problem will be to study the distribution of Z_n as $n \to \infty$.

Let P_X be the distribution of any random variable X on C, while P_Y represents the distribution on $[0, 1)$ of the corresponding Y. (Again $X = e^{2\pi i Y}$.) The analogue of the characteristic function of a real random variable is the *characteristic sequence,* defined as

$$\phi_X(m) = E(X^m) = \int_C z^m \, dP_X(z) = \int_0^1 e^{2\pi i m t} \, dP_Y(t), \tag{8}$$

where m is an integer. Of course, $\phi_X(m)$ is just the mth Fourier coefficient of the measure P_Y. It is easy to see that

$$\phi(0) = 1 \quad \text{and} \quad |\phi(m)| \leq 1 \quad \text{for all } m.$$

The multiplicative property of the expectations of independent random variables (Theorem 3.1) extends without difficulty to complex-valued variables, and so the characteristic sequence for Z_n is the nth power of the common characteristic sequence of the X_k. This fact is the key to understanding the limiting distribution of Z_n.

To answer our question about the distribution of Z_n, some tools are needed. These results are analogous to the theorems of Section 15 concerning characteristic functions, but the compactness of the circle makes matters simpler. We will leave most of the preliminary work to the reader.

Lemma 2 (Uniqueness Theorem). *Two probability measures on C with the same characteristic sequence are identical.*

Problem 1. *Prove this lemma. [Hint: If the measures (call them P and Q) are different, there must be (why?) some continuous function f on C such that*

$$\int_C f(z)\,dP \neq \int_C f(z)\,dQ.$$

Now approximate this function by a trigonometric polynomial.]

Lemma 3 (Continuity Theorem). *Let P_n be probability measures on C with corresponding characteristic sequences $\phi_n(m)$, and assume that*

$$\lim_{n\to\infty} \phi_n(m) = \phi(m) \qquad \text{exists} \tag{9}$$

for each integer m. Then $\phi(m)$ is the characteristic sequence of a measure P on C, and $P_n \Rightarrow P$.

Problem 2. *Prove Lemma 3. [Hint: On a compact space such as C, any sequence of probability measures must have a weakly convergent subsequence.]*

With these preparations out of the way, it is easy to prove the following.

Theorem 2. *Provided that the common distribution of the X_k is not supported on a finite set of points that are equally spaced around the circle, the probability distribution of Z_n converges weakly to the uniform distribution on C.*

Proof. As noted already, the characteristic sequence of Z_n is

$$\phi_n(m) = E(Z_n^m) = \phi(m)^n \tag{10}$$

where $\phi(m)$ is the common characteristic sequence of the variables X_k. For all values of m such that $|\phi(m)| < 1$, the limit of (10) obviously will be 0; therefore in case

$$|\phi(m)| < 1 \qquad \text{for all } m \text{ except } m = 0 \tag{11}$$

we have

$$\lim_{n\to\infty} \phi_n(m) = \delta_{0,m} = \begin{cases} 1 & \text{for } m = 0, \\ 0 & \text{for } m \neq 0. \end{cases} \tag{12}$$

The right-hand side of (12) is the characteristic sequence of the uniform distribution, so Lemma 2 assures us that under these conditions the conclusion of the theorem does hold.

To interpret assumption (11), suppose it fails; then there is some $m_0 \neq 0$ such that $|\phi(m_0)| = 1$. This means that $\phi(m_0) = e^{2\pi i \theta}$ for some real θ, so that

$$1 = e^{-2\pi i \theta} \phi(m_0) = \int_0^1 e^{2\pi i (m_0 t - \theta)} dP_Y(t).$$

It follows that $\cos[2\pi(m_0 t - \theta)] = 1$ must hold almost everywhere with respect to P_Y; in other words, P_Y is supported on the set

$$S = \{t \in [0, 1) : (m_0 t - \theta) \text{ is an integer}\}.$$

The points $z = e^{2\pi i t}$ corresponding to the m_0 elements of S are therefore equally spaced around C as asserted in the theorem. □

Remark 1. What if the distribution of the X_k is supported on a finite set of equally spaced points? Suppose m_0 is the smallest number of points that will do. If these points are the m_0th roots of unity, then $\phi(m_0) = 1$, while $|\phi(m)| < 1$ for values of m not divisible by m_0. In this case, there is again a limiting measure, the uniform measure on the m_0th roots of unity. If the m_0 points supporting P_X are not roots of unity, they must be a translation of such a set of roots; this case occurs when $|\phi(m_0)| = 1$ but $\phi(m_0) \neq 1$. In this situation, the distribution of Z_n does not have a limit.

Problem 3. *Prove the assertions of this remark.*

Remark 2. It is suggestive to interpret these facts in group-theoretic terms. In the "general" case described in Theorem 2, the distribution of Z_n converges to the *Haar measure* (or the invariant measure) of the group C. If the distribution of the X_k is supported on a *compact subgroup* (the m_0th roots of unity), then the distribution of Z_n converges to the Haar measure of that subgroup. Finally, when the distribution of X_k lives on a *coset* of some compact subgroup, then (and only then) the distribution of Z_n has no limit. A question for further study (beyond the scope of this book) is: When and how do these facts generalize?

We return to the real line and convergence to the normal distribution. Even in this case, there is a great deal more to do. Questions about the rate of convergence, or the case of unequal distributions for the summands X_k, are important for applications and have received much attention. It is also possible to investigate other modes of convergence than the weak limits of Section 13; one example is the possible convergence of the density function of the normalized sum S_n to the normal density. We will mention, without proof, one major result of each type. For the complete story and proofs, see Gnedenko and Kolmogorov (1968) or Feller (1966).

If the normal approximation is going to be used in applications, it is important to have some bounds on the errors that are made. A major result in this direction is due to Cramer, Berry, and Esseen:

Theorem 3. *Let $X_1, X_2, \ldots$ be a sequence of independent random variables with the same distribution, and suppose that distribution has mean 0, variance σ^2, and finite third moment $M = E(|X_k^3|)$. Then there exists a constant $C \leq 3$ such that for every $x \in (-\infty, \infty)$,*

$$\left| P\left(\frac{S_n}{\sqrt{n}} \leq x\right) - N_\sigma(x) \right| \leq \frac{CM}{\sigma^3 \sqrt{n}}. \tag{13}$$

(The smallest value of C which works for all distributions is not known.) If higher moments than the third also exist, the error is still of the order of $n^{-1/2}$ but it can be calculated more explicitly.

For the case of unequal distributions, the definitive result was obtained in 1922 by J. W. Lindeberg.

Theorem 4. *Let $X_1, X_2, \ldots$ be a sequence of independent random variables with distributions $P_1, P_2, \ldots$. Assume that P_k has mean 0 and variance σ_k^2, and put $B_n^2 = \sigma_1^2 + \cdots + \sigma_n^2 = \mathrm{var}(S_n)$. Assume that for each $t > 0$,*

$$\lim_{n \to \infty} \frac{1}{B_n^2} \sum_{k=1}^n \int_{|u| \geq tB_n} u^2 \, dP_k(u) = 0. \tag{14}$$

Then the distributions of the sums S_n satisfy

$$P\left(\frac{X_1 + \cdots + X_n}{B_n} \leq x\right) \Rightarrow N(x). \tag{15}$$

Equation (14) is known as the *Lindeberg condition*; it was shown by W. Feller that it is necessary as well as sufficient for (15). In the identically distributed case, (14) reduces to

$$\lim_{n\to\infty} \int_{|u|\geq t\sqrt{n}} u^2\, dP(u) = 0$$

where P is the common distribution; this holds since the variance is finite.

Problem 4. *Show that if the random variables $X_1, X_2, \ldots$ have means 0, variances σ_k^2 that are bounded away from 0 and from ∞, and bounded third absolute moments, the Lindeberg condition is satisfied.*

Finally, we turn to the possible convergence of the density of S_n, suitably rescaled, to the normal density; results of this sort are called *local limit theorems*. Such convergence does not follow from the mere existence of a density for the X_k, even in the identically distributed case. However, the following is true.

Theorem 5. *Let $X_1, X_2, \ldots$ be a sequence of independent random variables with a common distribution having mean 0 and variance σ^2. Suppose this distribution is absolutely continuous with a bounded density. Then*

$$\lim_{n\to\infty} \sup_{-\infty < x < \infty} \left| \frac{d}{dx} P\left(\frac{S_n}{\sqrt{n}} \leq x \right) - \frac{1}{\sqrt{2\pi}\sigma} e^{-x^2/2\sigma^2} \right| = 0; \qquad (16)$$

that is, the density of $S_n/\sqrt{n}$ converges uniformly to the density of the normal distribution with mean 0 and variance σ^2.

There are also local limit theorems for the situation where the X_k take only integer values (the "lattice" case); in fact, the de Moivre limit theorem was first proved in this way. The goal now is an estimate for the probability that S_n takes on a particular value. In the case of lattice distributions, characteristic functions become Fourier series and the desired probabilities are particular coefficients of these series. A general theorem along these lines can be found in Gnedenko and Kolmogorov (1968).

17. STABLE DISTRIBUTIONS

For more than 200 years after the work of de Moivre, the study of limiting distributions in probability theory mostly meant finding conditions for convergence to the normal. (The principal exception, Poisson's limit the-

STABLE DISTRIBUTIONS

orem, seemed to be an isolated special case.) But in the twentieth century new worlds have been discovered; a pioneer in expanding the concept of limit theorems was the French mathematician Paul Lévy. In the next three sections we will survey this terrain, starting here with some suggestive examples and definitions.

The normal distribution has the remarkable property that the sum of two or more independent, normally distributed random variables is again normal, differing only by a change of scale. There are other distributions with this property. One such is the distribution of Cauchy (see Section 15, Problem 5) which has the characteristic function $\phi(\lambda) = \exp(-c|\lambda|)$, where c is a positive constant. The average S_k/k of k independent Cauchy variables has the same distribution as the individual terms since

$$E(e^{i\lambda S_k/k}) = \phi_{S_k}(\lambda/k) = (e^{-c|\lambda/k|})^k = e^{-c|\lambda|}. \tag{1}$$

We will now exhibit some more examples.

In 1920, the physicist Holtzmark raised and solved the following problem: If electrically charged particles are located "randomly" in space, what will be the probability distribution of the resulting electric field at a fixed point? About 1940, Chandrasekhar studied analogous problems concerning the gravitational field resulting from a random distribution of stars. Here is a simplified, one dimensional model in the spirit of these investigations.

Assume that n "stars" (points) are distributed randomly in the interval $[-n, n]$. Their locations $X_1, \ldots, X_n$ are independent and uniformly distributed in this interval. Each star has mass $m > 0$ and the gravitational constant is unity. Then the force that will be exerted on a unit mass at the origin (the field strength) is

$$Y_n = \sum_{k=1}^{n} \frac{m \, \text{sign}(X_k)}{X_k^2}. \tag{2}$$

We will show that *the distributions of the random variables Y_n have a weak limit as $n \to \infty$.*

The approach, naturally, is based on characteristic functions. First, we use the assumption that X_k is uniformly distributed to compute

$$E(e^{i\lambda m \, \text{sign}(X_k)/X_k^2}) = \int_{-n}^{n} e^{i\lambda m \, \text{sign}(x)/x^2} \frac{1}{2n} \, dx = \frac{1}{n} \int_0^n \cos\left(\frac{\lambda m}{x^2}\right) dx.$$

Because the X_k are independent, the characteristic function of Y_n is just the nth power of this expression. Making some simple changes, we can pass to the limit:

$$E(e^{i\lambda Y_n}) = \left(1 - \frac{1}{n}\int_0^n \left[1 - \cos\left(\frac{\lambda m}{x^2}\right)\right] dx\right)^n$$

$$= \left\{1 - \frac{1}{n}\int_0^\infty \left[1 - \cos\left(\frac{\lambda m}{x^2}\right)\right] dx + o\left(\frac{1}{n}\right)\right\}^n$$

$$\to \exp\left(-\int_0^\infty \left[1 - \cos\left(\frac{\lambda m}{x^2}\right)\right] dx\right).$$

If we make a change of variable in the integral, our result can finally be written as

$$\lim_{n\to\infty} E(e^{i\lambda Y_n}) = e^{-c|\lambda|^{1/2}}, \tag{3}$$

where $c > 0$ is constant. Since the limit is continuous, by Theorem 15.3 it must be the characteristic function of a probability distribution to which the distributions of the Y_n converge weakly. It is not possible to express the distribution function of this limiting measure in elementary terms, although a good deal is known about its properties.

Problem 1. *Suppose that the inverse-square attraction in (2) is replaced by an inverse pth-power attraction. Show that for $p > 1/2$, (3) will be replaced by*

$$\lim_{n\to\infty} E(e^{i\lambda Y_n}) = e^{-c|\lambda|^{1/p}}, \qquad c > 0. \tag{4}$$

The distributions we have just discovered share with the Cauchy and the normal the *self-reproducing* property mentioned above. Suppose that $X_1 + \cdots + X_k = S_k$ denotes the sum of k independent random variables, each with the characteristic function (4) (and the same value of c). Then it is evident that S_k has that same characteristic function except that c must be replaced by kc. It follows just as for the Cauchy case that S_k/k^p has the same distribution as the individual X's.

We formalize these ideas with two definitions. First, suppose that the random variables X and Y have distribution functions F and G, respectively. If the variables are related by a linear change of scale

$$Y = aX + b, \qquad a > 0 \quad \text{and} \quad b \quad \text{constants,}$$

then, as noted in Section 2, their distribution functions must satisfy

$$G(t) = F\left(\frac{t-b}{a}\right) \quad \text{for all } t. \tag{5}$$

Definition 1. *Two distribution functions related by (5) for constants $a > 0$ and b are said to be* of the same type; *we indicate this by writing $F \sim G$. Two probability measures on R^1 are of the same type when (5) holds for the corresponding distribution functions.*

Problem 2. *Show that "being of the same type" is an equivalence relation between distributions. (That is, show that it is reflexive, symmetric, and transitive.)*

The set of all distributions on R^1 is thus divided into equivalence classes or "types." Note that all normal distributions form one equivalence class and all degenerate distributions[7] form another. Using this terminology, we can restate the self-reproducing property:

Definition 2. *Suppose that $S_k = X_1 + \cdots + X_k$ denotes the sum of k independent random variables, each with the same nondegenerate distribution P. The distribution P is said to be* stable *if the distribution of S_k is of the same type as P for every positive integer k. A random variable is called stable if its distribution has this property.*

Problem 3. *Prove that stability is a property of equivalence classes; that is, if a distribution P is stable, so is any distribution of the same type as P.*

Explicitly, Definition 2 requires that there must exist constants $a_k > 0$ and b_k such that

$$P\left(\frac{S_k - b_k}{a_k} \leq t\right) = P(X_1 \leq t) \quad \text{for all } k > 1. \tag{6}$$

Equivalently, the characteristic function of P must satisfy

$$e^{-i\lambda b_k/a_k} \phi(\lambda/a_k)^k = \phi(\lambda) \quad \text{for all } k > 1. \tag{7}$$

---

[7]"Degenerate" distributions are those that concentrate all their probability at a single point.

The characteristic functions $\exp(-c|\lambda|^{1/p})$ in (4) satisfy (7) with $b_k = 0$ and $a_k = k^p$; $1/p$ is called the *index* of the distribution. If $p = 1$, we obtain the Cauchy distribution. The normal has $a_k = k^{1/2}$ and so its index is 2, although it is included in (4) only as a limiting case.[8]

We will next establish some properties of the stable distributions and try to see how large a class they form. The main result shows that we have already encountered all the symmetric stable laws on R^1.

Theorem 1. *Suppose that P is a stable distribution that is symmetric about $x = 0$. Then the characteristic function of P is given by*

$$\phi(\lambda) = e^{-c|\lambda|^\alpha}, \qquad (8)$$

where $c > 0$ and $\alpha \in (0, 2]$ are constants.

Proof. Let ϕ be the characteristic function of P. We will first show that $\phi(\lambda)$ is never 0 on the real line. From Corollary 15.2 we know that ϕ is real; it is also even and continuous, with $\phi(0) = 1$. Hence if ϕ vanishes somewhere, there must be a smallest positive number λ_0 such that $\phi(\lambda_0) = 0$. But the function ϕ satisfies (7) with all the $b_k = 0$ by symmetry, and so we have $\phi(\lambda) = \phi(d\lambda)^2$ for some positive d. It is impossible that $d = 1$, since this would make ϕ identically 1 and the distribution P would be degenerate. But $\lambda_0 d$ and λ_0/d must be zeros of ϕ along with λ_0, and so when $d \neq 1$ it is clear that no smallest positive 0 can exist. Consequently, $\phi(\lambda)$ is never 0 and so by continuity we have $\phi(\lambda) > 0$ for all real λ. Thus we can write $\psi(\lambda) = \log \phi(\lambda)$ without ambiguity. We know too that $\psi(\lambda)$ is real and nonpositive, the latter because characteristic functions are bounded by 1.

We need the following fact.

Lemma 1. *If $\psi(a\lambda) = \psi(b\lambda)$ for all $\lambda > 0$ where a and b are positive constants, then $a = b$.*

Proof. Suppose $a > b$. By iterating the hypothesis of the lemma, we see that

$$\psi\left(\left[\frac{b}{a}\right]^n \lambda\right) = \psi(\lambda).$$

---

[8]On the negative side, the Poisson distribution is not stable (even though the sum of independent Poisson random variables is itself Poisson), and neither are the other familiar distributions of elementary statistics.

STABLE DISTRIBUTIONS

Letting $n \to \infty$ and using continuity, we have $\psi(\lambda) = \psi(0) = 0$ for all λ, which means that the distribution P is degenerate. Hence, $a = b$.

We return to the proof of Theorem 1. The condition (7) for a characteristic function to be that of a stable distribution can now be stated in terms of ψ: For any positive integer k, there exists a constant $a_k > 0$ such that

$$k \psi(\lambda) = \psi(a_k \lambda). \tag{9}$$

Because of the lemma, the constants a_k are uniquely defined. Moreover,

$$\psi(a_{mn}\lambda) = mn \psi(\lambda) = m \psi(a_n \lambda) = \psi(a_m a_n \lambda),$$

so that from the lemma we must have

$$a_{mn} = a_m a_n \tag{10}$$

for all positive integers m and n. Our immediate goal will be to extend (9) and (10) from the integers to all positive real numbers.

It is convenient to rewrite a_k with the functional notation $A(k)$. Clearly $A(1) = 1$. We next define this function for positive rational numbers by setting $A(m/n) = A(m)/A(n)$. Because of property (10), it is easy to see that the value of $A(r)$ for any rational number is well defined, and that both (9) and (10) still hold when k, m, and n are any rationals. (Verify.)

Now suppose that a sequence of positive rationals tends to a positive limit; that is, $r_n \to x > 0$. We will prove that $\lim A(r_n)$ exists. First, notice that

$$\lim_{n \to \infty} \psi(A(r_n)\lambda) = \lim_{n \to \infty} r_n \psi(\lambda) = x \psi(\lambda). \tag{11}$$

It follows that the sequence $\{A(r_n)\}$ cannot have 0 as a limit point, for if it did the left-hand side of (11) would have to be $\psi(0) = 0$, which would again produce the degenerate case. The same trick exactly using the sequence $\{1/r_n\}$ shows that the sequence $\{A(r_n)\}$ also cannot have a subsequence tending to infinity. Since $\{A(r_n)\}$ is bounded away from both 0 and ∞, it must have a subsequence converging to a positive number u. But if another subsequence converges to v, from (11) we obtain $\psi(u\lambda) = \psi(v\lambda)$, and by the lemma $u = v$. Hence $\lim A(r_n) > 0$ exists. This limit is independent of the particular sequence $\{r_n\}$ (and thus depends only on x) since any two sequences of rationals converging to x could be interlaced and the new sequence $\{r_{n'}\}$ would also be one for which $\lim A(r_{n'})$ exists, impossible if the two limits were different. We therefore define $A(x) = \lim A(r_n)$, where the r_n are rational and converge to $x > 0$.

It is now easy to see that (9) and (10) hold for all positive x, becoming, respectively,

$$x \psi(\lambda) = \psi(A(x)\lambda) \qquad \text{for all } x > 0 \tag{12}$$

and

$$A(xy) = A(x) A(y) \qquad \text{for all } x, y > 0. \tag{13}$$

It is also not hard to verify that $A(x)$ is continuous; to see this, we just apply the argument used above for rationals to show that $A(x_n) \to A(x)$ whenever $x_n \to x$. (The reader can fill in the details.) Now we are nearly finished, for these facts imply that $A(x) = x^p$ for some constant p.

Problem 4. *Prove that the most general continuous solution of* (13) *is of the form* $A(x) = x^p$.

Remark. (13) is one of four functional equations, obtainable one from another by changing variables, which characterize the linear, power, exponential, and logarithmic functions. Some regularity condition is necessary in order to rule out "strange" (nonmeasurable) solutions; continuity and monotonicity are each sufficient.

Returning to Eq. (12) armed with the fact that $A(x) = x^p$, we substitute s for $A(x)$ and 1 for λ. The result is that for $s > 0$,

$$\psi(s) = \psi(1) s^{1/p}. \tag{14}$$

Since ψ is an even function, this establishes the form (8) of the stable characteristic function with $1/p = \alpha$ and $c = -\psi(1)$. The condition $c \geq 0$ is automatic; $c = 0$ would give a degenerate distribution and so we must have $c > 0$. For the same reason, $\alpha \neq 0$, while $\alpha < 0$ would contradict the fact that ϕ is continuous at the origin with $\phi(0) = 1$. It only remains to show that $\alpha \leq 2$ and the theorem will be proved.

The last step goes as follows: Suppose $\alpha > 2$; then it is easy to see that $\phi''(0) = 0$. As we have already remarked (without proof so far, but see below), the existence of $\phi''(0)$ implies the finiteness of the second moment, and since $\phi''(0) = 0$, that moment (i.e., the variance) has to vanish. This would correspond to another degenerate distribution—but such a distribution does not have a characteristic function of the form (8). It follows that if $\alpha > 2$, (8) is not a characteristic function at all. □

STABLE DISTRIBUTIONS

Problem 5. *Show that if a characteristic function ϕ is real and if $\phi''(0)$ exists, then so does $E(X^2)$. [Hint: Since ϕ is real, we can write*

$$-\frac{1}{2}\phi''(0) = \lim_{h \to 0} \int_{-\infty}^{\infty} \frac{1 - \cos hx}{h^2} \, dP;$$

Now apply Fatou's lemma.]

For symmetric stable distributions, the parameter α in (8) is the index, and the norming constants are given by $a_k = k^{1/\alpha}$. In fact, the constants a_k have this form (and so the concept of index applies) for *all* stable laws, not only symmetric ones.

Corollary 1. *For any stable distribution P on R^1, the norming constants $\{a_k\}$ are given by $a_k = k^{1/\alpha}$, where $0 < \alpha \leq 2$.*

Proof. The distribution P can be convolved with its reflection in the origin; the result is the symmetric distribution of the random variable $X - X'$, where X and X' are independent and each has the distribution P. It is easy to see that this convolution is stable whenever P is and has the same norming constants $\{a_k\}$. (Verify this.) Since we have proved the result for symmetric stable distributions, it must be true for them all. □

Problem 6. *Using the same symmetrization trick, show that the result in Problem 5 holds without assuming that the characteristic function is real.*

Corollary 2. *All stable distributions are absolutely continuous and have continuous densities.*

Proof. If ϕ is the characteristic function of a stable distribution, then $|\phi|^2$, which is the characteristic function of the random variable $X - X'$ mentioned above, is of the form $\exp(-c|\lambda|^\alpha)$. It follows that ϕ is integrable on R^1 for any $\alpha > 0$, and so the result follows from Corollary 15.3. □

We have found all the symmetric (and hence real-valued) stable characteristic functions, and we know the absolute value of *any* such function. How much is still missing? It is, of course, possible to translate any stable distribution, but that does not produce a new type. It turns out that when $\alpha < 2$ there is one additional parameter that measures the nonsymmetry of the distribution. (When $\alpha = 2$, we already know them all.) We will not derive the most general (nonsymmetric) stable c.f., but will be content to discuss one example—which, incidentally, is the third, and apparently the last, stable distribution whose density can be written in elementary terms. (The other two are the normal and the Cauchy.) This example shows the maximum possible nonsymmetry since it puts all the probability on the positive half-line.

Our approach is indirect. Suppose $T_1, T_2, \ldots$ are the times of the successive returns to the starting point in a simple random walk on the integers.[9] (We anticipate from Section 20 the fact that the walk returns repeatedly to 0 with probability 1.) We will prove the following.

Theorem 2. *For the simple random walk, the limiting distribution*

$$\lim_{n \to \infty} P\left(\frac{T_n}{n^2} \leq x\right) = F(x) \tag{15}$$

exists and satisfies

$$\int_0^\infty e^{-sx} dF(x) = e^{-\sqrt{2s}} \qquad \text{for } \Re(s) \geq 0. \tag{16}$$

The distribution F is stable with index $\alpha = 1/2$.

Proof. The time T_1 of first return is an integer-valued random variable, and we will show in Section 20 that it has the *generating function*

$$G(s) = E(s^{T_1}) = 1 - \sqrt{1 - s^2}. \tag{17}$$

But the times between successive returns are independent variables with this same distribution, since the walk can be thought of as "starting anew" upon each return. (This is an example of the *Markov property*, which we will discuss in Chapter 4.) Hence T_n is the sum of n independent copies of T_1, and so its generating function is $G(s)^n$. Making the change of variable $s = e^{-w}$, $w \geq 0$, and introducing the norming constants n^2, we obtain

$$E(e^{-wT_n/n^2}) = G(e^{-w/n^2})^n = [1 - \sqrt{1 - e^{-2w/n^2}}]^n.$$

If we pass to the limit, the result is

$$\lim_{k \to \infty} E(e^{-wT_n/n^2}) = e^{-\sqrt{2w}}, \qquad w \geq 0. \tag{18}$$

---

[9] Such a "random walk" consists of the partial sums of independent random variables, each taking the values $+1$ or -1 with probabilities $1/2$.

STABLE DISTRIBUTIONS

But by Theorem 15.5, (18) implies the weak convergence of distributions asserted in the theorem.

Finally, it is easy to show that the limiting distribution is stable by appealing to the uniqueness theorem for the Laplace transform. In fact, the sum of k independent variables with the distribution in question has the transform $\exp(-k\sqrt{2w})$; this is the same as the transform for one variable if w is replaced by w/k^2. This completes the proof of Theorem 2. □

Remark. There are elementary formulas for the distribution function $F(x)$ and for its density. In fact, as tables of Laplace transforms indicate,

$$\frac{1}{\sqrt{2\pi}} \int_0^\infty e^{-st} e^{-1/2t} t^{-3/2} \, dt = e^{-\sqrt{2s}} \tag{19}$$

so that the density of $F(x)$ is given by

$$f(x) = \frac{1}{\sqrt{2\pi}} e^{-1/2x} x^{-3/2} \tag{20}$$

for $x > 0$ [and $f(x) = 0$ for $x \leq 0$]. The distribution function itself is thus related to the normal distribution $N(x)$:

$$F(x) = 2\left[1 - N\left(\frac{1}{\sqrt{x}}\right)\right] \tag{21}$$

(again for $x > 0$). These formulas have elementary proofs, which we will leave as two last problems for the reader.

Problem 7. *In tables of integrals, the formula*

$$\int_0^\infty e^{-x^2} e^{-a^2/x^2} \, dx = \frac{\sqrt{\pi}}{2} e^{-2a} \tag{22}$$

can be found; derive this formula. [Hint: Denote the integral by $h(a)$; then find $h'(a)$ by differentiating under the integral sign. The resulting integral can be transformed by substitutions into a constant multiple of $h(a)$.]

Problem 8. *Using (22), obtain the Laplace transform formula (19).*

18. LIMIT DISTRIBUTIONS FOR SUMS AND MAXIMA

In Theorem 17.2, as in the central limit theorem, the limiting distribution for the partial sums of a sequence of random variables turns out to be stable. This is no accident. In this section we will show that the class of stable distributions coincides with the possible limiting distributions for the partial sums of independent, identically distributed (i.i.d.) random variables. The same method yields a characterization of limiting distributions for partial maxima of such a sequence; a property similar to stability is the key here as well. From this result, we will be able to show that the examples of limit distributions for maxima noted in Section 14 are the only ones possible.

First, some preparations. $F \star G$ will denote the convolution of distribution functions (as in Theorem 3.2), and $F^{\star k}$ means the k-fold convolution of F with itself (the kth "convolution power"). The corresponding characteristic function, naturally, is the ordinary kth power ϕ^k. We will need several lemmas.

Lemma 1. *Suppose that F_n, F, G_n, and G are all probability distribution functions, and that $F_n \Rightarrow F$ and $G_n \Rightarrow G$. Then $F_n \star G_n \Rightarrow F \star G$.*

Proof. Let ϕ_n, ϕ, ψ_n, and ψ, respectively, be the characteristic functions of F_n, F, G_n, and G; then $\phi_n \psi_n$ is the characteristic function of $F_n \star G_n$. We know that $\phi_n \to \phi$ and that $\psi_n \to \psi$; hence $\phi_n \psi_n \to \phi \psi$. But the product $\phi \psi$ is the characteristic function of $F \star G$. The conclusion therefore follows from Theorem 15.2 (the continuity theorem). □

Lemma 2. *Suppose that F_n and G are distribution functions, that $a_n > 0$ and b_n are constants, and that $F_n(a_n x + b_n) \Rightarrow G(x)$. Suppose that $\alpha_n > 0$ and β_n are other sequences of constants such that*

$$\lim_{n \to \infty} \frac{\alpha_n}{a_n} = 1 \quad \text{and} \quad \lim_{n \to \infty} \frac{\beta_n - b_n}{a_n} = 0. \tag{1}$$

Then it is also the case that $F_n(\alpha_n x + \beta_n) \Rightarrow G(x)$.

Problem 1. *Prove this lemma, using Theorem 13.1.*

The next result is the critical tool for showing stability.

Lemma 3. *Suppose that G_n, G, and H are distribution functions, and that*

$$G_n(x) \Rightarrow G(x) \quad \text{and} \quad G_n(a_n x + b_n) \Rightarrow H(x) \qquad (2)$$

for some constants $a_n > 0$ and b_n. Assume also that neither G nor H is degenerate. Then G and H must be of the same type.

Proof. We will first show that $\{a_n\}$ has a subsequence tending to a positive, finite limit a. If this is not so, there must be a subsequence $\{a_{n'}\}$ tending either to 0 or to ∞. Suppose for the moment that $b_n = 0$ for all n. Then if $a_{n'} \to 0$, using the assumption that $G_n \Rightarrow G$ we can conclude that for continuity points of H we must have

$$H(x) = \lim_{n \to \infty} G_n(a_n x) \begin{cases} \leq G(0+) & \text{for } x > 0; \\ \geq G(0-) & \text{for } x < 0. \end{cases}$$

This is impossible since the limit G is assumed to be nondegenerate. Similarly, if $a_{n'} \to \infty$ we would obtain $H(x) = 1$ for $x > 0$ and $H(x) = 0$ for $x < 0$, excluded because H is not degenerate. Thus we must have a subsequence such that $a_{n'} \to a$, where $0 < a < \infty$.

To remove the assumption that $b_n = 0$, we again use the symmetrization trick introduced in the last section. Given any distribution function F, let F^- be the distribution of the measure corresponding to F but reflected through the origin. [If F is the distribution function of a random variable X, then F^- is the d.f. of $-X$. Thus $F^-(t) = 1 - F(-t)$ at continuity points.] Clearly $G_n \Rightarrow G$ implies $G_n^- \Rightarrow G^-$ and $G_n(a_n x + b_n) \Rightarrow H(x)$ implies $G_n^-(a_n x - b_n) \Rightarrow H^-(x)$, so by Lemma 1

$$G_n \star G_n^- \Rightarrow G \star G^-$$

and

$$G_n(a_n x + b_n) \star G_n^-(a_n x - b_n) \Rightarrow H(x) \star H^-(x).$$

But it is easy to see that

$$G_n(a_n x + b_n) \star G_n^-(a_n x - b_n) = G_n(a_n x) \star G_n^-(a_n x)$$

since the terms $+b_n$ and $-b_n$ represent translations that are made in both directions and therefore cancel. Thus the distribution functions $G_n \star G_n^-$ satisfy the hypotheses of the lemma with no translation terms b_n and with the same scaling constants a_n that appear in (2). It follows from the paragraph above, then, that $\{a_n\}$ has a subsequence converging to a positive, finite limit.

We return now to (2), with n restricted to the sequence $\{n'\}$ for which $a_{n'} \to a > 0$. It is not hard to see that if some subsequence of $\{b_{n'}\}$ tends to either $+\infty$ or $-\infty$, then H must again be degenerate. We can therefore choose a subsubsequence $\{n''\}$ for which $b_{n''} \to b$ (finite), while, of course, $a_{n''} \to a$ holds as well. From the first part of (2), we then have

$$G_{n''}(ax + b) \Rightarrow G(ax + b),$$

while at the same time

$$G_{n''}(a_{n''}x + b_{n''}) \Rightarrow H(x).$$

It follows by Lemma 2 that $H(x) = G(ax + b)$, so that H and G are of the same type as asserted. □

Now the main results can be proved quite easily.

Theorem 1 (Lévy). *Let $X_1, X_2, \ldots$ be independent, identically distributed random variables with $S_n = X_1 + \cdots + X_n$, and suppose that there exist constants $a_n > 0$ and b_n such that*

$$F_n(x) = P\left(\frac{S_n - b_n}{a_n} \le x\right) \Rightarrow F(x) \tag{3}$$

for some nondegenerate distribution function F. Then F is stable.

Proof. We must show that $F^{\star k}$ is of the same type as F for each $k > 1$. By assumption $F_n \Rightarrow F$, and so from Lemma 1 we know that $F_n^{\star k} \Rightarrow F^{\star k}$. But the convolution $F_n^{\star k}$ is the distribution function of the sum of k independent random variables, each with the same distribution as $(S_n - b_n)/a_n$. Hence,

$$F_n^{\star k}(x) = P\left(\frac{S_n - b_n}{a_n} + \frac{S_{2n} - S_n - b_n}{a_n} + \cdots + \frac{S_{kn} - S_{(k-1)n} - b_n}{a_n} \le x\right)$$

$$= P\left(\frac{S_{kn} - kb_n}{a_n} \le x\right) = P\left(\frac{S_{kn} - b_{kn}}{a_{kn}} \le \frac{a_n x + kb_n - b_{kn}}{a_{kn}}\right)$$

$$= F_{kn}\left(\frac{a_n}{a_{kn}}x + \frac{kb_n - b_{kn}}{a_{kn}}\right). \tag{4}$$

But from (3) we know that $F_{kn}(x) \Rightarrow F(x)$. Hence Lemma 3 applies, with F_{kn} in the place of G_n, with F in the role of G, and with $F^{\star k}$ in place of H. The conclusion shows that F is stable. □

LIMIT DISTRIBUTIONS FOR SUMS AND MAXIMA   113

Remark. We have shown that any limit distribution in (3) must be stable. Can the possibilities be restricted further? The answer is no, and the proof is easy.

Problem 2. *If F is any stable distribution function, show that (3) will hold for suitable constants when each X_k has itself the distribution F.*

Theorem 1 answers one question but raises another: Under what conditions on the common distribution of the X_k will constants exist so that (3) holds? If such constants do exist, we say that the distribution belongs to the *domain of attraction* of the stable law F, or that the distribution of the X_k *is attracted to* F. For example, according to Theorem 16.1 every distribution with finite variance is attracted to the normal distribution. A complete answer to this question is given in Gnedenko and Kolmogorov (1968); we will be content to illustrate the situation in a special case, stated in the form of two problems.

Problem 3. *Let G be the distribution function of a probability measure that is symmetric about 0 and has the characteristic function ϕ. Suppose that*

$$\lim_{x \to \infty} x[1 - G(x)] = c, \qquad c > 0. \tag{5}$$

Prove that $\phi(\lambda) = 1 - c\pi|\lambda| + o(|\lambda|)$ as $\lambda \to 0$. [Hint: Write

$$\frac{1 - \phi(\lambda)}{2\lambda} = \int_0^\infty \frac{1 - \cos \lambda x}{\lambda} \, dG(x)$$

and break up the integral into parts that can each be estimated.]

Problem 4. *Show that the distribution function G of the preceding problem is attracted to the Cauchy distribution (see Problem 15.5).*[10]

Turning our attention to the maximum term in a random sequence, we can characterize the possible limiting distributions in much the same way we handled the corresponding problem for sums. Parallel to the concept of stability (Definition 17.2) we have the following.

---

[10] In fact, (5) is close to being necessary as well as sufficient. The precise condition for a symmetric distribution to be attracted to the Cauchy law is that $1 - G(x) = x^{-1}L(x)$, where L is a slowly varying function (defined in Section 14). These functions play an important role in the theory of attraction.

Definition 1. *Suppose that $M_k = \max(X_1, X_2, \ldots, X_k)$ denotes the maximum of k independent random variables, each with the same nondegenerate distribution P. This distribution (and the corresponding distribution function F) is said to be* maximal *if the distribution of M_k is of the same type as P for every positive integer k.*[11]

The distribution function of the maximum of several independent random variables is just the product of their individual distribution functions [recall (14.2)]. Therefore, the condition for a distribution to be maximal is that F^k should be of the same type as F for every k. This is formally the same as stability for sums, but with the ordinary kth power in place of the k-fold convolution. The relationship with limiting distributions is also the same as for sums.

Theorem 2 (Fisher and Tippett; Gnedenko). *Let $X_1, X_2, \ldots$ be independent, identically distributed random variables with $M_n = \max(X_1, \ldots, X_n)$, and suppose that there exist constants $a_n > 0$ and b_n such that*

$$F_n(x) = P\left(\frac{M_n - b_n}{a_n} \le x\right) \Rightarrow F(x) \tag{6}$$

for some nondegenerate distribution function F. Then F is maximal.

Proof. A small piece of notation will be useful: $M_{i,j}$, $i < j$, will mean the maximum of the random variables $X_{i+1}, \ldots, X_j$. (Thus $M_n = M_{0,n}$.) To prove the theorem, we must show that for each $k > 1$, F^k is of the same type as F. Again by assumption $F_n \Rightarrow F$, and so clearly $F_n^k \Rightarrow F^k$. But the power F_n^k is the distribution function of the maximum of k independent random variables, each with the same distribution as $(M_n - b_n)/a_n$. Hence,

$$F_n(x)^k = P\left(\max\left[\frac{M_{0,n} - b_n}{a_n}, \frac{M_{n,2n} - b_n}{a_n}, \ldots, \frac{M_{(k-1)n, kn} - b_n}{a_n}\right] \le x\right)$$

$$= P\left(\frac{M_{kn} - b_n}{a_n} \le x\right) = P\left(\frac{M_{kn} - b_{kn}}{a_{kn}} \le \frac{a_n x + b_n - b_{kn}}{a_{kn}}\right)$$

$$= F_{kn}\left(\frac{a_n}{a_{kn}} x + \frac{b_n - b_{kn}}{a_{kn}}\right). \tag{7}$$

---

[11] The reader is warned that (unlike the definition of a "stable" distribution) this use of the word "maximal" is not standard. The term "max-stable" is sometimes used for this concept.

But from (6) we know that $F_{kn}(x) \Rightarrow F(x)$. Therefore Lemma 3 applies just as before and shows that F and F^k are of the same type. This proves that F is maximal. □

Finally, we will use Theorem 2 to determine all possible maximal distributions, and so all the limiting distributions for a sequence of partial maxima. This solves completely the problem whose analog for sums—determining all the stable distributions—was only partially resolved in Section 17.

Theorem 3. *Every maximal distribution is of the same type as one of the limiting distributions* Φ_α, Ψ_α, *or* Λ *defined in equations* (14.5), (14.7), *and* (14.9), *respectively*.

Proof. Let F be a maximal distribution function; this means that for each integer $k > 1$ there exist constants $\alpha_k > 0$ and β_k such that

$$F^k(\alpha_k t + \beta_k) = F(t). \tag{8}$$

Let us first suppose that there is some value of k such that $\alpha_k > 1$. This leads to one of the three types of solution; as we will see, the other two solutions arise when there is an $\alpha_k < 1$ and when $\alpha_k = 1$ for all k.

Because $\alpha_k > 1$, we have

$$\alpha_k t + \beta_k < t \quad \text{when} \quad t < \frac{-\beta_k}{\alpha_k - 1} = r_k, \tag{9}$$

and the same thing holds with both inequalities reversed. Thus for $t < r_k$, since F is monotonic we have

$$F(t) \geq F(\alpha_k t + \beta_k) \geq F^k(\alpha_k t + \beta_k) = F(t). \tag{10}$$

The inequalities thus must be equalities so that $F(t) = 0$ or 1, and since (10) holds for all $t < r_k$, the value must be 0. It is also easy to see that $F(r_k) = 0$ or 1; we can rule out 1 since the distribution F would be degenerate in that case. Hence $F(t) = 0$ for all $t \leq r_k$.

We will show now that $0 < F(t) < 1$ for every $t > r_k$. In fact, if $F(t_0) = 0$ for some $t_0 > r_k$, then from (8) we see that $F(\alpha_k t_0 + \beta_k) = 0$ as well, where $\alpha_k t_0 + \beta_k > t_0$. When this is iterated, we find that $F(t) = 0$ for arbitrarily large values of t, which is impossible for a distribution function. Thus, $F(t) = 0$ holds if and only if $t \leq r_k$. Moreover, if $F(t) = 1$ holds for some t, there must be a smallest such value by right-continuity. Call this smallest value t_1; clearly $t_1 > r_k$. Now choose t_2 so that $\alpha_k t_2 + \beta_k = t_1$ and substitute t_2 for t in (8); we see that $F(t_2) = 1$ also. But $t_2 < t_1$, which contradicts the choice of t_1 as the least such value. Thus $F(t) < 1$ for all t. To summarize, we have shown that $F(t) = 0$ for $t \leq r_k$ and that $0 < F(t) < 1$ for $t > r_k$.

The next step is to verify that $\alpha_k > 1$ holds for every $k > 1$ whenever it is true for one such k. Suppose that $\alpha_m < 1$ for some $m > 1$. Then, by an argument entirely similar to that above, we could show that $F(t) = 1$ for all sufficiently large t, which would contradict the form of F already established. Similarly, if $\alpha_n = 1$ for some $n > 1$, it is easy to show that $F(t)$ never takes the values 0 or 1. Thus either of these possibilities is inconsistent with the assumption that there exists $k > 1$ for which $\alpha_k > 1$.

It is now clear that the values of r_k, which represent the largest point where $F(t) = 0$, must be the same for all $k > 1$; let us call this common value r. [Thus $r = -\beta_k/(\alpha_k - 1)$ for any $k > 1$.] Define $G(t) = F(t + r)$. Then obviously G is a distribution function of the same type as F, with $G(t) = 0$ for all $t \leq 0$ and with $0 < G(t) < 1$ for $t > 0$. It is easy to check that the maximality condition (8) transforms into

$$G^k(\alpha_k t) = G(t) \qquad \text{for } k = 2, 3, \ldots. \tag{11}$$

It remains to find all the solutions of (11).

To proceed, we use a change of variable. For $t > 0$ let $\psi(t) = \log G(t)$; then $-\infty < \psi(t) < 0$ and

$$k\psi(\alpha_k t) = \psi(t).$$

This is essentially the same as Eq. (17.9) which we encountered in our study of stable distributions, and its solution is very similar. The reader can verify that the argument given in Section 17 is valid in the present case, with the result that

$$\alpha_k = k^p \qquad \text{and} \qquad \psi(s) = \psi(1) s^{-1/p};$$

Since $\alpha_k > 1$, we know that $p > 0$. Hence, for $t > 0$

$$G(t) = \exp(-ct^{-1/p}) \tag{12}$$

where c and p are positive constants; we know that $G(t) = 0$ for $t \leq 0$. From this, it is clear that G, and so also F, is of the same type as the distribution $\Phi_\alpha(t)$ given in (14.5), with $\alpha = 1/p$. This completes the proof in the case when $\alpha_k > 1$ for some $k > 1$.

Now suppose that, for some $k > 1$, we have $\alpha_k < 1$. Almost exactly the same reasoning leads to the conclusion that $F(t)$ must then be of the same type as one of the distributions $\Psi_\alpha(t)$ defined in (14.7); we omit the details which closely follow the argument above.

Finally, we must consider the case when $\alpha_k = 1$ for some, and hence for all, $k > 1$. As we have noted above, in this case $0 < F(t) < 1$ for all t. The maximality condition (8) becomes $F^k(t + \beta_k) = F(t)$ and so $\beta_k > 0$. We make these changes of variables:

$$e^{\beta_k} = c_k, \qquad e^t = z, \qquad F(t) = G(z). \tag{13}$$

Then G is a distribution function with $G(0+) = 0$, $0 < G(z) < 1$ for $z > 0$, and satisfying

$$G^k(c_k z) = G(z) \qquad \text{for } z > 0. \tag{14}$$

This is exactly the same as (11) and so its solution is

$$G(z) = \exp(-cz^{-1/p}) \qquad \text{for } z > 0.$$

Reversing the changes of variable (13), we obtain

$$F(t) = \exp(-ce^{-(1/p)t}), \tag{15}$$

which is of the same type as $\Lambda(t)$, the limiting distribution in (14.9). This completes the proof of Theorem 3. □

Remarks. Continuing the analogy with sums of random variables and stable distributions, we have the problem of determining the *domain of maximal attraction* for each of the possible limiting distributions. The simple results in Section 14 give some idea what to expect, and Gnedenko solved this problem rather completely in the 1943 paper cited earlier. We will not take these matters further in this book.

It is easy to envision extending the problem of limiting distributions for sums of independent random variables from one to several dimensions, but it seems less obvious how the theory of the maximum of a random sample should be generalized. One approach is to construct the *convex hull* of a collection of n independent random vectors in R^k, and to investigate its limiting behavior as $n \to \infty$. Some interesting results in this direction have been found,[12] but the problem is far from completely solved. [See also Galambos (1987), Chapter 5.]

19. INFINITELY DIVISIBLE DISTRIBUTIONS

Once the central limit problem was freed from the restriction to normal convergence still more general questions could be asked and answered. Moreover, one familiar limit theorem—that of Poisson (see Theorem 15.6)—was still not part of the picture. The most obvious generalization of stable convergence is to ask for the possible limits in (18.3) when the summands X_k no longer need to have the same distribution. There is one caution if this problem is to be meaningful: We must ensure that X_{n+1} is small compared to S_n as $n \to \infty$, for without some such condition anything can happen. With a suitable restriction of this sort, the problem is a good one, and it has been solved. The possible limits are called *L-distributions,* and they form a much larger class than the family of stable laws. The Poisson distribution, however, is not yet included.

It turns out to be fruitful to generalize still further and consider the limits of *triangular arrays* of random variables $X_k^{(n)}$, where $k = 1, 2, \ldots, N(n)$. For each n, we will assume that $X_1^{(n)}, \ldots, X_{N(n)}^{(n)}$ are independent and identically distributed (i.i.d.). Let $n \to \infty$. We now ask: If $N(n) \to \infty$ and if

$$F_n(x) = P(X_1^{(n)} + \cdots + X_{N(n)}^{(n)} \leq x) \Rightarrow F(x), \qquad (1)$$

which nondegenerate distributions F can appear? This question includes the convergence studied in Section 18, since if $X_1, X_2, \ldots$ is a sequence of identically distributed variables we can choose $N(n) = n$ and define

$$X_k^{(n)} = \frac{1}{a_n}\left(X_k - \frac{b_n}{n}\right);$$

then the distribution F_n of (1) is the same as F_n in (18.3). Poisson's limit theorem is included too, as seen by taking again $N(n) = n$ and letting the $X_k^{(n)}$ equal 1 or 0 according to the outcomes of Bernoulli trials with success probability equal to μ/n. In this case, F will be the Poisson distribution with parameter μ. It is easy to see that this distribution is not stable.

---

[12] See, for example, Lloyd Fisher, "Limiting sets and convex hulls of samples from product measures," *Annals Math. Stat.* 40 (1969), pp. 1824–1832.

INFINITELY DIVISIBLE DISTRIBUTIONS

Definition 1. *A distribution function F is said to be* infinitely divisible *if F has a kth root with respect to convolution—that is, F is the kth convolution power of some other distribution function G_k—for every $k = 2, 3, \ldots$.*

We say that a probability distribution is infinitely divisible if the corresponding distribution function satisfies Definition 1. It is equivalent to require that the characteristic function of F is the kth power of some other characteristic function for every $k = 2, 3, \ldots$. For example, in the case of the Poisson distribution we have

$$\phi(\lambda) = e^{-\mu} \sum_{n=0}^{\infty} \frac{\mu^n}{n!} e^{i\lambda n} = \exp(\mu e^{i\lambda} - \mu) \tag{2}$$

which is indeed the kth power of a characteristic function, another Poisson with parameter μ/k. Similarly, a standard normal distribution is the kth convolution power of a normal with mean 0 and $\sigma = 1/\sqrt{k}$. Both the Poisson and the normal, therefore, are infinitely divisible.

Problem 1. *Show that every stable law is infinitely divisible.*

Problem 2. *Prove that if F is infinitely divisible, so is any other distribution of the same type as F.*

Problem 3. *Prove that the* exponential distribution *defined by*

$$F(x) = \begin{cases} 1 - e^{-ax} & \text{for } x > 0, \\ 0 & \text{for } x \leq 0 \end{cases} \tag{3}$$

($a > 0$) is infinitely divisible and find its "kth root." Is F stable?

The first goal of this section is to show that the infinitely divisible distributions are just the possible limits that can appear in (1). We will prove this under a slight simplifying assumption.

Theorem 1. *Suppose that $N(n) = n$, that $X_1^{(n)}, \ldots, X_n^{(n)}$ are independent with a common distribution for each n, and that (1) holds. Then F is infinitely divisible.*

Proof. To show that F has a kth convolution root, consider the subsequence $\{F_{n'}\}$ where $n' = mk$, $m = 1, 2, \ldots$. Group the random variables making up the sum in (1) into k blocks, so that

$$S_{n'} = X_1^{(mk)} + \cdots + X_{mk}^{(mk)} = Y_1^{(m)} + \cdots + Y_k^{(m)},$$

where

$$Y_j^{(m)} = X_{(j-1)m+1}^{(mk)} + \cdots + X_{jm}^{(mk)},$$

for $j = 1, \ldots, k$. The k random variables $Y_1^{(m)}, \ldots, Y_k^{(m)}$ are then independent and identically distributed; let $G_m(x) = P(Y_j^{(m)} \leq x)$ be their common distribution function. If the distributions $\{G_m\}$ converge to a weak limit as $m \to \infty$, the kth convolution power of that limit will equal the distribution F by Lemma 18.1. The same is true even if only a subsequence of $\{G_m\}$ converges to a limit. Therefore to prove the theorem it suffices to show that the distributions $\{G_m\}$ form a conditionally compact family, since in that case a convergent subsequence must exist.

Suppose the sequence $\{G_m\}$ is not tight; specifically, that condition (13.20) does not hold. (Recall Theorem 13.3.) Then by definition there exists some $\epsilon > 0$ such that for every constant A, we have either

$$G_m(-A) \geq \epsilon \quad \text{or} \quad 1 - G_m(A) \geq \epsilon \qquad (4)$$

for at least one value of m. In fact, (4) must hold for infinitely many m, since in the contrary case a larger value of A could be chosen for which the inequalities were never true. At least one of the two inequalities in (4) then must hold infinitely often; suppose it is the second one. In this case, we have

$$P(Y_1^{(m)} + \cdots + Y_k^{(m)} > kA) \geq \epsilon^k$$

for arbitrarily large m, so that by (1) the limit F satisfies $1 - F(kA) \geq \epsilon^k$. Since this holds for all A, F cannot be a distribution function. Similarly, if it is the first case that holds infinitely often, we obtain $F(-kA) \geq \epsilon^k$ for every A and again F cannot be a distribution. This contradiction shows that $\{G_m\}$ must be tight, and so it has a weakly convergent subsequence. It follows that F has a kth root. This works for any k, and so F is infinitely divisible. □

Remark 1. The converse of Theorem 1—the fact that every infinitely divisible distribution can arise as a limit of the form (1) with $N(n) = n$—is immediate from the definition of infinite divisibility. (Why?)

INFINITELY DIVISIBLE DISTRIBUTIONS

Remark 2. The conditions in Theorem 1 can be relaxed; for example, the assumption that $N(n) = n$ could be replaced by $N(n) \to \infty$. Moreover, the $X_1^{(n)}, \ldots, X_{N(n)}^{(n)}$ need not be identically distributed, provided that a condition is imposed to ensure that each individual term is small compared to the sum. These changes do not enlarge the class of possible limits; F still must be infinitely divisible.

Two natural questions confront us. First, just what *are* the infinitely divisible (i.d.) distributions? Second (analogous to the problem of attraction), when does (1) hold for a particular i.d. distribution F? We will give a partial answer to the first of these and omit the second one altogether; the whole story can be found in Gnedenko and Kolmogorov (1968). The results in this area are remarkably complete.

Example. Suppose that $X_1, X_2, \ldots$ are independent, identically distributed random variables, and let N be a Poisson-distributed variable independent of the X_k. The distribution of the *random sum*

$$S_N = \sum_{k=1}^{N} X_k \tag{5}$$

is then said to be *compound Poisson*. If we use characteristic functions and the "conditioning" technique (recall Example 2 of Section 4), it is easy to see that all such distributions are infinitely divisible:

$$\psi(\lambda) = E(e^{i\lambda S_N}) = \sum_{n=0}^{\infty} E(e^{i\lambda S_N} | N = n) P(N = n)$$

$$= \sum_{n=0}^{\infty} E(e^{i\lambda(X_1 + \cdots + X_n)}) P(N = n)$$

$$= \sum_{n=0}^{\infty} \phi(\lambda)^n e^{-\mu} \frac{\mu^n}{n!} = \exp(\mu\phi(\lambda) - \mu), \tag{6}$$

where ϕ denotes the characteristic function of the X_k. But if we replace μ by μ/k, the same setup produces a characteristic function whose kth power is just $\psi(\lambda)$. This shows that the distribution of S_N is infinitely divisible.

It is a little surprising that this simple scheme comes close to yielding all infinitely divisible laws, in the following sense.

Theorem 2. *A distribution P is infinitely divisible if and only if P is the weak limit of a sequence of compound-Poisson distributions.*

Proof. The "if" part of the theorem is an immediate consequence of the infinite divisibility of all compound-Poisson distributions and the following.

Lemma 1. *The weak limit of a sequence of infinitely divisible distributions is itself infinitely divisible.*

Proof. Let the convergent sequence of distributions be $\{P_n\}$. Since P_n is infinitely divisible, it has an nth root Q_n; let us imagine a triangular array of random variables as in Theorem 1 where Q_n serves as the distribution of $X_j^{(n)}$. Then P_n will be the distribution of $S_n = X_1^{(n)} + \cdots + X_n^{(n)}$. By the conclusion of Theorem 1, the limit of the distributions P_n must be infinitely divisible.

The proof of the "only if" part of Theorem 2 requires some preparation.

Lemma 2. *The characteristic function of an infinitely divisible distribution never vanishes.*

Proof. For any real characteristic function $\psi(\lambda)$, we have

$$1 - \psi(2\lambda) \leq 4[1 - \psi(\lambda)]. \tag{7}$$

This fact is easy to prove. Let $G(x)$ be the distribution function corresponding to the characteristic function ψ; since ψ is real, the corresponding measure must be symmetric about 0 by Corollary 15.2. Then

$$1 - \psi(2\lambda) = \int_{-\infty}^{\infty} (1 - \cos 2\lambda x)\, dG(x) = 2\int_{-\infty}^{\infty} \sin^2 \lambda x\, dG(x)$$

$$= 2\int_{-\infty}^{\infty} (1 - \cos \lambda x)(1 + \cos \lambda x)\, dG(x)$$

$$\leq 4\int_{-\infty}^{\infty} (1 - \cos \lambda x)\, dG(x) = 4[1 - \psi(\lambda)].$$

Now suppose that ϕ is the infinitely divisible characteristic function in question. We can assume ϕ is real, since in any case $|\phi|^2$ is a real characteristic function, also infinitely divisible (why?) and with the same set of zeros as ϕ. Since ϕ is continuous and $\phi(0) = 1$, there is an interval, say $[-a, a]$, on which $\phi(\lambda) > 0$.

If ϕ_k is that kth root of ϕ which is itself a characteristic function, it is easy to see from continuity that $\phi_k(\lambda) > 0$ also, at least for all $\lambda \in [-a, a]$. It follows that $\phi_k(\lambda) \to 1$ uniformly on that closed interval as $k \to \infty$, and so for large enough k, we must have $1 - \phi_k(\lambda) < 1/4$ for all $\lambda \in [-a, a]$. By (7), we then have $1 - \phi_k(\lambda) < 1$ for $\lambda \in [-2a, 2a]$ so that ϕ_k cannot vanish over this larger range. The same must hold for its kth power, which is the characteristic function ϕ.

We have shown that if ϕ does not vanish on an interval $[-a, a]$, it cannot vanish on $[-2a, 2a]$. It follows that ϕ does not vanish anywhere, as asserted.

Lemma 3. *Let ϕ be an infinitely divisible characteristic function, and for each k, let ϕ_k be a characteristic function whose kth power is ϕ. Then*

$$\lim_{k \to \infty} \phi_k(\lambda) = 1 \qquad \text{for all real } \lambda. \tag{8}$$

Proof. Since $\phi(0) = 1$ while $\phi(\lambda)$ is continuous and never vanishes, there is a unique way to define the function $\arg \phi(\lambda)$ that is continuous and makes $\arg \phi(0) = 0$. There is similarly a unique definition of $\arg \phi_k(\lambda)$ meeting those requirements, and in fact we must have simply

$$\arg \phi_k(\lambda) = \frac{\arg \phi(\lambda)}{k}.$$

To see this, note that since $\phi_k(\lambda)^k = \phi(\lambda) \neq 0$, the possibilities for $\arg \phi_k(\lambda)$ are

$$\arg \phi_k(\lambda) = \frac{\arg \phi(\lambda)}{k} + j\frac{2\pi}{k}, \qquad j = 0, 1, \ldots, k - 1.$$

At $\lambda = 0$ the choice $j = 0$ is required, and since $\arg \phi(\lambda)$ and $\arg \phi_k(\lambda)$ are continuous, it is impossible to switch to any value of j other than 0. Given all this we can write

$$\phi_k(\lambda) = |\phi(\lambda)|^{1/k} e^{i \arg[\phi(\lambda)]/k},$$

from which the limit (8) is immediate.

Returning to the proof of Theorem 2, let ϕ be the characteristic function of the infinitely divisible distribution P. We will use logarithms of the characteristic functions ϕ and ϕ_k; they are defined with that version of arg ϕ discussed above. We have

$$\log \phi(\lambda) = k \log \phi_k(\lambda),$$

and because of Lemma 3, we can approximate $\log \phi_k(\lambda)$ by $\phi_k(\lambda) - 1$. The result is

$$\log \phi(\lambda) = \lim_{k \to \infty} k [\phi_k(\lambda) - 1],$$

which is the same as

$$\phi(\lambda) = \lim_{k \to \infty} e^{k \phi_k(\lambda) - k}. \qquad (9)$$

The exponential on the right side of (9) is the characteristic function of a compound-Poisson distribution for each k. By the continuity theorem, the limit relation (9) is equivalent to the weak convergence of these distributions to P, which proves Theorem 2. □

Problem 4. *Exhibit a sequence of compound-Poisson distributions that converges to the standard normal distribution. [Note: Following the proof above gives one way to accomplish this. What are some other ways?]*

Theorem 2 describes the class of all infinitely divisible distributions, but specific representations for these distributions are also known. Kolmogorov discovered a formula for the characteristic function of any i.d. distribution with finite variance, and a little later Khintchine and Lévy extended that result to the most general case. We will state Kolmogorov's result here and invite the reader to prove it by solving the four problems that follow.

Theorem 3. *Suppose that $\phi(\lambda)$ is the characteristic function of an infinitely divisible distribution with mean μ and (finite) variance σ^2. Then*

$$\phi(\lambda) = \exp \left(i\lambda\mu + \int_{-\infty}^{\infty} \frac{e^{i\lambda x} - 1 - i\lambda x}{x^2} \, dG(x) \right), \qquad (10)$$

where G is a nondecreasing function of bounded variation.

Problem 5. *If ϕ is the characteristic function of an i.d. distribution function F with mean μ, show that*

$$\log \phi(\lambda) = i\mu\lambda + \lim_{k \to \infty} \int_{-\infty}^{\infty} (e^{i\lambda x} - 1 - i\lambda x) k \, dF_k(x), \qquad (11)$$

where F_k is the distribution function whose k-fold convolution is F.

Problem 6. *If in addition F has finite variance σ^2, prove that the monotonic functions*

$$G_k(x) = k \int_{-\infty}^{x} u^2 \, dF_k(u) \tag{12}$$

are uniformly bounded. Show that consequently there is a subsequence $\{G_{k'}\}$ whose corresponding measures converge weakly to a limit G.

Problem 7. *For the subsequence $\{G_{k'}\}$ of the previous problem, show that*

$$\lim_{k' \to \infty} \int_{-\infty}^{\infty} \frac{e^{i\lambda x} - 1 - i\lambda x}{x^2} \, dG_{k'}(x) = \int_{-\infty}^{\infty} \frac{e^{i\lambda x} - 1 - i\lambda x}{x^2} \, dG(x), \tag{13}$$

where G is nondecreasing and of total variation at most σ^2.

Combining (11), (12), and (13) establishes Kolmogorov's formula (10). We leave the converse as one more problem.

Problem 8. *Show that (10) defines the characteristic function of an infinitely divisible distribution for every nondecreasing function G of bounded total variation. [Hint: Use Theorem 2.]*

We end this section with one final question for the reader.

Problem 9. *What is the representation (10) (i.e., find the function G) for the characteristic function of a normal distribution with mean μ and variance σ^2?*

20. RECURRENCE

We have completed our survey of limiting distributions for sums (and maxima) of independent random variables; generally speaking, these problems were completely solved during the 1930s and early 1940s. However, this theory does not contain everything worth knowing about sequences of independent variables. In Chapter 2 we studied properties of the sequence $\{S_n\}$ of partial sums that hold (or fail) almost surely, and here we will consider one more question of that kind: whether or not the partial sums return to zero (or at least approach close to zero) infinitely often. The variables will always be assumed to have a common distribution.

A *simple random walk* in d dimensions means the sequence of partial sums of independent random vectors, each of which takes as its possible values the $2d$ unit vectors along the coordinate axes (in both the positive and the negative directions) choosen with equal probabilities $1/2d$. Equivalently, we can think of the walk as a Markov chain[13] whose states are the lattice points of R^d and whose transitions consist of moves from a lattice point to one of its $2d$ immediate neighbors, selected with equal probabilities $1/2d$. As with any infinite-state Markov chain, it is basic to ask whether or not the random walk is *recurrent* in the sense that it returns to its starting state with probability 1.[14] This question was answered in 1921 by George Polya, who showed that the answer is *yes* for $d = 1$ and 2, but *no* for all $d \geq 3$ [see Feller (1968)]. Thirty years later Polya's interesting discovery was generalized by K. L. Chung and W. H. J. Fuchs, who replaced the simple random walk by the partial sums of general independent, identically distributed random variables or vectors.[15] We will discuss here only the one-dimensional case, and follow Chung and Fuchs by using characteristic functions to analyze the problem. (Other methods are now known.)

For technical simplicity, we will assume that the random values take integer values. (Chung and Fuchs did not use this restriction.) In this case, we call the sequence $\{S_n\}$ of partial sums *recurrent* provided that

$$P(S_n = 0 \text{ for infinitely many } n) = 1; \qquad (1)$$

otherwise, the process is said to be *transient*. [As we will soon see, the probability in (1) must then be 0.] Notice that if the random variables being summed have a continuous distribution, then $P(S_n = 0) = 0$ for $n > 0$ and recurrence as defined in (1) is impossible. For this reason, in the general case Chung and Fuchs study not recurrence but *interval recurrence*, meaning that $\{S_n\}$ returns almost surely to any neighborhood of 0 rather than to 0 itself. With this change, the results in the general case are much the same as for integer-valued variables.

We start by noting one fact that is a corollary of previous results: *If $E(X_k) = \mu$ exists and $\mu \neq 0$, then $\{S_n\}$ is transient.* This follows from the strong law of large numbers (Theorem 9.4), since under these conditions $S_n/n \to \mu \neq 0$ holds almost surely and so $\{S_n\}$ tends to either $+\infty$ or $-\infty$,

---

[13] Knowledge of Markov chains is not necessary for reading this section.
[14] The term *persistent* is also used for this property.
[15] "On the distribution of values of sums of random variables," *Memoirs Amer. Math. Soc.* 6 (1951).

depending on the sign of μ. On the other hand, we will prove that *if $\mu = 0$, then $\{S_n\}$ is recurrent*. Notice that this does not follow directly from the law of large numbers, or even from such stronger results as the law of the iterated logarithm.[16]

It is useful to introduce some notation. Let $u_n = P(S_n = 0)$ be the probability of a return to the origin after n steps, and put $u_0 = 1$. Let f_n be the probability that the *first* return occurs on the nth step; that is,

$$f_n = P(S_1 \neq 0, \ldots, S_{n-1} \neq 0, S_n = 0). \tag{2}$$

Of course, $\sum_{n=1}^{\infty} f_n = f \leq 1$ represents the probability that a return to the origin ever occurs. It is easy to see from these definitions that

$$u_n = \sum_{k=1}^{n} f_k u_{n-k} + \delta_{n,0}, \tag{3}$$

where $\delta_{n,m} = 1$ if $n = m$ and 0 otherwise. Equation (3) can be solved recursively for either sequence $\{u_n\}$ or $\{f_n\}$ if the other is known, so the two sequences determine each other.

A convenient way to exploit relation (3) is to introduce the *generating functions* of $\{u_n\}$ and $\{f_n\}$, which we denote by $U(x)$ and $F(x)$ respectively.[17] Taking the generating function of both sides of (3), we find

$$U(x) = F(x)U(x) + 1$$

which is easily solved for either F or U; in particular,

$$U(x) = \frac{1}{1 - F(x)}. \tag{4}$$

Lemma 1. *The sequence of partial sums $\{S_n\}$ is recurrent if and only if*

$$\sum_{n=0}^{\infty} u_n = \sum_{n=0}^{\infty} P(S_n = 0) = \infty. \tag{5}$$

---

[16] There are exceptions to this negative statement. The law of the iterated logarithm, when it holds, implies that $\{S_n\}$ changes sign infinitely often. If the only positive value taken on by the X_k is $+1$, then a change of sign from $S_n < 0$ to $S_{n+m} > 0$ can only happen if $S_{n+k} = 0$ for some $k < m$. (A corresponding statement holds if -1 is the only possible negative value for X_k.) The simple random walk, in particular, satisfies both conditions and its recurrence does follow from the law of the iterated logarithm.

[17] That is, $U(x) = \sum_{n=0}^{\infty} u_n x^n$ and $F(x) = \sum_{n=1}^{\infty} f_n x^n$.

Proof. If $\sum u_n < \infty$, then the Borel–Cantelli lemma asserts that with probability 1 only finitely many of the events $\{S_n = 0\}$ occur, so that the process is transient. If the sum is infinite, by letting $x \to 1-$ in (4), we see that $F(1) = 1$. This means that $\sum f_n = 1$ so that one return to the origin is certain. But $F(x)^k$ is the generating function for the probabilities of making the kth return on the nth step; these probabilities also sum to 1. Therefore k returns to the origin are also certain to occur, for any k. This proves recurrence. $\square$

Example. For the simple random walk in one dimension, the choices of moving to the left or to the right are Bernoulli trials with $p = 1/2$. Since a return to the origin in exactly $2n$ steps requires that the walk take n steps in each direction, while a return in an odd number of steps is impossible, we have

$$P(S_{2n} = 0) = \binom{2n}{n}\left(\frac{1}{2}\right)^{2n}, \qquad P(S_{2n+1} = 0) = 0.$$

The generating function $U(x)$ can now be calculated:

$$U(x) = \sum_{n=0}^{\infty} \binom{2n}{n}\left(\frac{x}{2}\right)^{2n} = \frac{1}{\sqrt{1-x^2}}. \tag{6}$$

From (4) we then obtain

$$F(x) = 1 - \sqrt{1-x^2}, \tag{7}$$

a result that was needed in Section 17. Since $F(1) = 1$ or $U(1) = \infty$, the random walk is recurrent.

Problem 1. *Derive the second part of (6). [Hint: Use the binomial theorem.]*

Returning to the general (integer-valued) case, suppose that ϕ is the common characteristic function of the random variables X_k; then

$$\phi(\lambda)^n = E(e^{i\lambda S_n}) = \sum_{k=-\infty}^{\infty} e^{i\lambda k} P(S_n = k).$$

This is a Fourier series, and so the inversion formula we need is simply the definition of the 0th Fourier coefficient:

$$P(S_n = 0) = \frac{1}{2\pi} \int_{-\pi}^{\pi} \phi(\lambda)^n \, d\lambda.$$

RECURRENCE

It is tempting to sum under the integral sign, but some care is needed. It is easy, however, to justify the following:

$$\sum_{n=0}^{\infty} P(S_n = 0) x^n = \frac{1}{2\pi} \int_{-\pi}^{\pi} \frac{1}{1 - x\phi(\lambda)} d\lambda \qquad \text{for } |x| < 1. \qquad (8)$$

Combining (8) with the lemma above yields the following.

Theorem 1. $\{S_n\}$ *is recurrent or transient according to whether*

$$L = \lim_{x \to 1-} \frac{1}{2\pi} \int_{-\pi}^{\pi} \frac{1}{1 - x\phi(\lambda)} d\lambda \qquad (9)$$

is infinite or finite.

We will use this result to establish the fact mentioned earlier.

Theorem 2. *If $E(X) = 0$, then $\{S_n\}$ is recurrent.*

Proof. From the first lemma of Section 16, we know that $\phi'(\lambda)$ exists and is continuous, with $\phi'(0) = 0$. Thus for any $\epsilon > 0$, it is possible to choose a $\delta > 0$ such that

$$|1 - \phi(\lambda)| \leq \epsilon |\lambda| \qquad \text{for } |\lambda| \leq \delta. \qquad (10)$$

The quantity $1 - \phi(\lambda)$ need not be real, but (10) implies that both its real and imaginary parts are at most $\epsilon|\lambda|$ in the neighborhood we have chosen.

To use the estimate (10), we first observe that for $0 \leq x < 1$,

$$\int_{-\pi}^{\pi} \frac{d\lambda}{1 - x\phi(\lambda)} = \int_{-\pi}^{\pi} \Re\left[\frac{1}{1 - x\phi(\lambda)}\right] d\lambda$$

$$\geq \int_{-\delta}^{\delta} \Re\left[\frac{1}{1 - x\phi(\lambda)}\right] d\lambda = \int_{-\delta}^{\delta} \frac{1 - x\Re\phi(\lambda)}{|1 - x\phi(\lambda)|^2} d\lambda.$$

But from (10) we see that when $|\lambda| \leq \delta$,

$$|1 - x\phi(\lambda)|^2 = |(1 - x) + x(1 - \phi(\lambda))|^2 \leq [|1 - x| + x|1 - \phi(\lambda)|]^2$$

$$\leq [1 - x + x\epsilon|\lambda|]^2 \leq 2(1 - x)^2 + 2\epsilon^2|\lambda|^2.$$

Substituting, we obtain

$$\int_{-\pi}^{\pi} \frac{d\lambda}{1 - x\phi(\lambda)} \geq \frac{1}{2} \int_{-\delta}^{\delta} \frac{1 - x}{(1 - x)^2 + \epsilon^2 \lambda^2} d\lambda = \frac{1}{\epsilon} \arctan \frac{\delta\epsilon}{1 - x}.$$

Thus it is clear that

$$\liminf_{x\to 1-} \int_{-\pi}^{\pi} \frac{d\lambda}{1 - x\phi(\lambda)} \geq \frac{\pi}{2\epsilon},$$

and since $\epsilon > 0$ was arbitrary, this means that the limit in (9) is infinite. Hence according to Theorem 1, $\{S_n\}$ is recurrent. □

The existence of $E(X_k)$ is not a necessary condition for recurrence, and by methods similar to the above recurrence can sometimes be proved without it. Here is an example.

Problem 2. *Suppose the random variables $\{X_k\}$ take integer values and have a common distribution symmetric about 0 with distribution function G satisfying*

$$\lim_{x\to\infty} x[1 - G(x)] = c, \qquad c > 0.$$

Show that $E(X_k)$ does not exist and the weak law of large numbers does not hold, but $\{S_n\}$ is nevertheless recurrent. [Hint: Use the results of problems 18.3 and 18.4.]

The reader may be interested in the facts for higher dimensions; they resemble Polya's discovery about simple random walks. In the plane $E(X_k) = 0$ is no longer by itself sufficient for recurrence, but if in addition the second moments are finite, then recurrence again holds. (Of course, if the mean exists but is not 0, the process will be transient because of the strong law of large numbers.) When the dimension is 3 or more, recurrence is impossible as long as the distribution of the random vectors $\{X_k\}$ is not supported in a two-dimensional subspace. (These facts are due to Chung and Fuchs.) Recurrence and many other aspects of random walks were discussed extensively by F. Spitzer (1964) using quite different methods. We will not go further with these matters here.

CHAPTER FOUR

The Brownian Motion Process

21. BROWNIAN MOTION

We begin in the middle of things, with an important example that illustrates much of what the theory of stochastic processes is all about. Under certain conditions, small particles suspended in a fluid can be observed to undergo a continual, irregular motion. This phenomenon is named after its nineteenth century discoverer, Robert Brown. The explanation for the motion lies in the fact that the particles undergo innumerable collisions with "randomly" moving molecules in the surrounding fluid. Each collision individually has a negligible effect on the much heavier particle, but cumulatively they produce the observable motion.

A probabilistic theory for the Brownian motion, based on several simple assumptions, was put forward by Albert Einstein in 1906.[1] Let x_t denote one coordinate of the Brownian particle at time t, where for reference we take $x_0 = 0$. The underlying molecular motions are known only statistically and so we treat the position coordinate x_t as a random variable. The net displacement of the particle during any time interval (s, t) will be the sum of a vast number of small, approximately independent contributions resulting from individual molecular impacts, so in view of the central limit theorem, it is reasonable to expect the increment $x_t - x_s$ to be normally distributed. If the surrounding fluid has high enough viscosity, any velocity acquired by the particle will be damped out almost instantly; as a result, we can assume that the displacements in nonoverlapping intervals of time

---

[1] The French mathematician L. Bachelier, working on a theory for fluctuations in stockmarket prices, suggested much the same mathematical model at about the same time. Today, the term "Brownian motion" often refers to the mathematical object rather than the physical phenomenon.

will be independent. Finally, if the physical conditions are isotropic and constant in time and space, we can expect that $E(x_t) = 0$ and that the variance $E([x_{t+s} - x_t]^2) = f(s)$ of any increment should be independent of t. This condition, plus the assumed independence, implies that $f(t+s) = f(t) + f(s)$ so that $f(t) = ct$ for some constant $c > 0$.[2]

Problem 1. *Prove the assertion that* $\operatorname{var}(x_{t+s} - x_t) = cs$.

These considerations already form a theory that can be tested experimentally, but from a mathematical point of view they leave much to be desired. The model was more completely described by Norbert Wiener in 1923, and his work leads us to state the following.

Definition 1. *A standard Brownian motion process (or Wiener process) means a family* $\{x_t(\omega) : t \geq 0\}$ *of real-valued random variables defined on some probability space* $(\Omega, \mathcal{B}, P)$ *that satisfies:*

1. $x_0 = 0$.
2. *If* $0 = t_0 < t_1 < t_2 < \cdots < t_n$, *then the random variables* $x_{t_k} - x_{t_{k-1}}$ *for* $k = 1, 2, \ldots, n$ *are independent.*[3]
3. *For each* $s, t \geq 0$, *the random variable* $x_{t+s} - x_t$ *is normally distributed with mean 0 and variance* s.
4. *For almost all* $\omega \in \Omega$, *the function* $x_t = x_t(\omega)$ *is everywhere continuous in* t.

A "process"—that is, a family of random variables—satisfying this definition would fulfill the desiderata of Einstein's theory, but the existence of such a family is not at all obvious. (Of course the chief difficulty lies with point 4.)[4] Much of this chapter can be understood by simply assuming that an object satisfying Definition 1 is given. However, the existence question is itself of much interest and taking the shortcut is not recommended.

There are several methods for constructing the Wiener process (and many other random processes as well), and we are going to discuss three of them. The first approach, the only one we will carry out in full detail,

---

[2] In the mathematical theory the constant c has no particular significance and is often set equal to 1. However, this constant is important in the physical application. Einstein derived a relation between c and Avogadro's number which led to a new way to measure the latter by observing particles undergoing Brownian motion.
[3] A random process satisfying this condition is said to have *independent increments*.
[4] At this time, it may be useful to review the discussion of existence questions in Section 5.

BROWNIAN MOTION

begins with trying to represent the function x_t by an expansion into some sort of Fourier-like series. This reduces the noncountable set of random variables $\{x_t(\omega)\}$ to a countable number of (random) coefficients, and if we choose the right kind of series, these coefficients may even turn out to be independent. This line of attack seems promising and we will follow it for a while—until it runs into a serious technical difficulty. At that point, we switch to a modified version due to Ciesielski and Kampé de Feriet that is easier to handle.

Let us begin, then, with a family of functions $\{\psi_n(t)\}$ that form an orthonormal basis for L_2 over the unit interval. We assume that the Brownian motion process exists and attempt to express it by a generalized Fourier series of the form

$$x_t(\omega) = \sum_{n=1}^{\infty} a_n(\omega) \psi_n(t) \tag{1}$$

for $0 \leq t \leq 1$. Since x_t is (by assumption) a continuous function and therefore belongs to L_2, the coefficients are given by the Fourier formula

$$a_n(\omega) = \int_0^1 x_s(\omega) \psi_n(s) \, ds. \tag{2}$$

This exhibits a_n as a limit of linear combinations of normally distributed random variables, and so a_n must itself be normal. Taking the expectation under the integral sign, we also have $E(a_n) = 0$.

What about the covariances of the $\{a_n\}$? From Eq. (2) we have

$$\begin{aligned} E(a_n a_m) &= E\left(\int_0^1 x_t \psi_n(t) \, dt \int_0^1 x_s \psi_m(s) \, ds \right) \\ &= E\left(\int_0^1 \int_0^1 x_t x_s \psi_n(t) \psi_m(s) \, dt \, ds \right) \\ &= \int_0^1 \int_0^1 E(x_t x_s) \psi_n(t) \psi_m(s) \, dt \, ds \\ &= \int_0^1 \int_0^1 \min(s, t) \psi_n(t) \psi_m(s) \, dt \, ds. \end{aligned} \tag{3}$$

Problem 2. *Verify that $E(x_t x_s) = \min(t, s)$ for any $s, t \geq 0$.*

Equation (3) should hold for any orthonormal basis $\{\psi_n(t)\}$. But, as noted above, it will be very convenient if the coefficients $\{a_n\}$ turn out to be independent random variables. In that case they must be uncorrelated, so that $E(a_n a_m) = 0$ whenever $n \neq m$. To achieve this, we choose the basis $\{\psi_n\}$ to be the *eigenfunctions* of the integral operator with kernal $E(x_t x_s) = \min(t, s)$; that is, we have to find and use the nonzero solutions of the equation

$$\int_0^1 \min(t, s)\, \psi(s)\, ds = \lambda \psi(t). \tag{4}$$

These turn out to be

$$\psi_n(t) = \sqrt{2} \sin\left[\left(n + \frac{1}{2}\right)\pi t\right], \quad \text{where} \quad \lambda_n = \frac{1}{\pi^2(n + 1/2)^2}. \tag{5}$$

(The factor $\sqrt{2}$ is a normalizing constant.) Combining these choices with (2), we find that the coefficients $\{a_n\}$ should be independent, normal random variables with variances λ_n, so that (1) becomes

$$x_t(\omega) = \frac{\sqrt{2}}{\pi} \sum_{n=1}^{\infty} \frac{Z_n(\omega)}{n + 1/2} \sin\left[\left(n + \frac{1}{2}\right)\pi t\right], \tag{6}$$

where as usual the $\{Z_n\}$ are independent random variables with the standard normal distribution.

Problem 3. *Fill in the derivation of Eqs. (5) and (6).*

Where does all this leave us? If a process $\{x_t\}$ that satisfies the conditions of Definition 1 does exist, then the series (6), considered as a function of t, must converge to x_t in the L_2 sense for every ω such that $x_t(\omega)$ is continuous—and hence for almost all ω. Our hope is to turn the argument around and use (6) to show the existence of the Brownian motion process, based on the known existence of a sequence $\{Z_n\}$ of independent normal random variables. It is easy enough to see that the series in (6) converges a.s. for each $t \in [0, 1]$; for this we need only appeal to Theorem 9.2 or to the "three-series theorem" of Section 10. But to establish the continuity in t we need uniform convergence, and nothing we have done so far is much help there.[5]

---

[5] Norbert Wiener used a similar Fourier series approach when he gave the first rigorous theory of the (mathematical) Brownian motion. He then proved the a.s. uniform convergence of a subsequence of the partial sums of his series, which of course was sufficient to establish the continuity of the limit with respect to t.

We are not going to tackle this problem head on; instead, we use a variation on the theme developed above that makes the uniform convergence much easier to prove. What follows is meant to motivate this variation. The trick is to start with the derivative $x'_t = dx_t/dt$ of the Brownian motion instead of with x_t itself. Of course this poses some new difficulties, since the limit of $(x_{t+h} - x_t)/h$ as $h \to 0$ does not exist, either pointwise or in the L_2 mean. (Verify the latter statement.[6]) Nevertheless, proceeding formally as if the derivative did exist, we see that x'_t ought to be a normal random variable with mean 0, and that for $t \neq s$ the variables x'_t and x'_s should be independent. (Why?)

The covariance function $K(t, s) = E(x'_t x'_s)$ is a little trickier. By independence we should have $K(t, s) = 0$ when $t \neq s$, but for $t = s$ the value is infinite. However, we can approximate the derivatives by difference quotients and define

$$K_h(t, s) = E\left(\frac{x_{t+h} - x_t}{h} \frac{x_{s+h} - x_s}{h}\right), \qquad h > 0. \tag{7}$$

Then it is not hard to see that for any continuous function f and any $t > 0$ and $\epsilon > 0$, we must have

$$\lim_{h \to 0} \int_{t-\epsilon}^{t+\epsilon} K_h(t, s) f(s)\, ds = f(t). \tag{8}$$

We thus can think of the covariance K as a "function" (actually an operator) characterized by the property that

$$\int_{t-\epsilon}^{t+\epsilon} K(t, s) f(s)\, ds = f(t), \qquad t > 0 \text{ and } \epsilon > 0, \tag{9}$$

for all continuous functions f. Equation (9) identifies the covariance with the "Dirac delta function" of quantum physics, the best known example of a generalized function.

Problem 4. *Prove* (8).

We return to the previous line of argument, using the covariance "function" K satisfying (9). Substituting in (3), we see that *any* orthogonal basis $\{\psi_n\}$ will yield the simplest possible covariances for the coefficients $\{a_n\}$, namely, $E(a_n a_m) = 0$ when $n \neq m$ and $E(a_n^2) = 1$. Thus for any such basis

---

[6]The derivative does exist as a "generalized function" in the sense of L. Schwarz and others, but we will not use any such theory in our development.

the analog of (6) is the formal expansion

$$\frac{dx_t}{dt} = \sum_{n=1}^{\infty} Z_n(\omega)\, \psi_n(t), \tag{10}$$

where the $\{Z_n\}$ are independent standard normals as before.

The series (10), of course, does not converge. But we are really seeking an expression not for x_t' but for x_t, so we formally integrate and obtain

$$x_t(\omega) = \sum_{n=1}^{\infty} Z_n(\omega) \int_0^t \psi_n(u)\, du = \sum_{n=1}^{\infty} Z_n(\omega)\, \Psi_n(t). \tag{11}$$

It is this series (11) that we will use to show the existence of the Brownian motion process. The integrated functions $\Psi_n(t)$ are continuous, and so if the series converges uniformly in t, the sum must be continuous also. The great advantage of (11) over (6) is that the functions $\{\psi_n(t)\}$ can now be *any* complete orthonormal set, so we can make a choice that facilitates proving the uniform convergence.[7] The only disadvantage of (11) is that the integrated functions $\{\Psi_n(t)\}$ are no longer orthogonal, and this does not matter for the proof of existence.

So much for motivation. We will choose a convenient basis and actually carry out the construction in the next section.

22. THE FIRST CONSTRUCTION

Which basis should we use? The trigonometric functions or the exponentials $\{e^{2\pi i n t}\}$ are not very good choices, because with any of these the series (21.11) has only a factor n in the denominator, as does (21.6), and the uniform convergence is again difficult to establish. It turns out that the *Haar functions* work very well; these functions have nice properties that make the proof of uniform convergence easy. They are defined on $[0, 1]$ as follows: $H_0(t) = 1$ and

$$H_1(t) = \begin{cases} +1 & \text{for } 0 \leq t \leq \dfrac{1}{2}; \\ -1 & \text{for } \dfrac{1}{2} < t \leq 1, \end{cases} \tag{1}$$

---

[7] It has even been shown by M. Nisio that the series (11) is a.s. uniformly convergent for every orthonormal basis $\{\psi_n\}$. The proof of this nice result needs some advanced tools.

THE FIRST CONSTRUCTION 137

Then when k satisfies $2^n \leq k < 2^{n+1}$ we set

$$H_k(t) = \begin{cases} 2^{n/2} & \text{for } \dfrac{k-2^n}{2^n} \leq t \leq \dfrac{k-2^n+1/2}{2^n}; \\ -2^{n/2} & \text{for } \dfrac{k-2^n+1/2}{2^n} < t \leq \dfrac{k-2^n+1}{2^n}; \\ 0 & \text{otherwise.} \end{cases} \quad (2)$$

The integrals of the Haar functions, namely,

$$S_k(t) = \int_0^t H_k(u)\,du, \quad (3)$$

are called *Schauder functions*.

It is easy to check that the Haar functions are indeed orthonormal. They are also complete, and in fact they have the special property that the Fourier expansion of any continuous function converges uniformly to that function. We will not actually need that result, however; completeness will enter by means of Parseval's relation—true for any orthonormal basis—which states that inner products can be computed by summing products of Fourier coefficients:

$$\int_0^1 f(t)g(t)\,dt = (f,g) = \sum_{k=0}^{\infty}(f,H_k)(g,H_k), \quad (4)$$

where f and g are real, square-integrable functions on $[0,1]$.

The proof that the Haar functions are complete is very simple and we will give it (for completeness?); we'll show that a function orthogonal to all the Haar functions must vanish a.e. Suppose, then, that $f \in L_2[0,1]$ and satisfies

$$(f,H_k) = \int_0^1 f(t)H_k(t)\,dt = 0 \quad (5)$$

for every k. Let F be the indefinite integral of f:

$$F(t) = \int_0^t f(u)\,du.$$

Since $f \in L_2[0,1]$, we know that F exists and is continuous; obviously $F(0) = 0$. The equation $(f,H_0) = 0$ translates directly into $F(1) = 0$. Next, note that $(f,H_1) = 2F(1/2) - F(1)$; using assumption (5), we conclude that $F(1/2) = 0$. The fact that f is also orthogonal to H_2 then yields $F(1/4) = 0$, and its orthogonality to H_3 adds the fact that $F(3/4) = 0$. Continuing by mathematical induction, we find that $F(k/2^n) = 0$ for every dyadic rational number in $[0, 1]$. Since F is continuous, it must therefore vanish identically, and so $f = 0$ a.e. This proves that the Haar functions are indeed complete.

Problem 1. *Carry out the induction proof to show that $F(k/2^n) = 0$ for all n and $k < 2^n$.*

The principal goal of this section is the following.

Theorem 1. *Let $Z_0, Z_1, \ldots$ be a sequence of independent, standard normal random variables*[8] *and let $\{S_k\}$ be the Schauder functions. Then the random series*

$$x_t(\omega) = \sum_0^\infty Z_k(\omega) S_k(t) \qquad (6)$$

converges uniformly in t with probability 1, and the random variables $\{x_t\}$ form a Brownian motion process in the sense of Definition 21.1, for $0 \le t \le 1$.

Proof. The proof falls into two parts. First, we must demonstrate the asserted convergence of the series; this depends on particular properties of the Haar functions. Given the convergence, we then will prove that the function $x_t(\omega)$ defined by (6) has the properties of Brownian motion. The second part uses only Parseval's relation, and so the argument would work for any complete, orthonormal basis. This part also uses some facts about multidimensional normal distributions that are discussed in a postscript.

For the first stage, some preparations are needed.

Lemma 1. *A series of Schauder functions*

$$\sum_{k=0}^\infty a_k S_k(t) = s(t) \qquad (7)$$

converges uniformly on $[0, 1]$ provided that $|a_k| = O(k^\epsilon)$ for some $\epsilon < 1/2$.

[8] More explicitly, each Z_k has a normal distribution with mean 0 and variance 1.

THE FIRST CONSTRUCTION

Proof. The functions $S_k(t)$ are nonnegative, and for $2^n \leq k < 2^{n+1}$ they attain a maximum value of $2^{-(n/2)-1}$. Furthermore, as k varies over this range the functions S_k have disjoint supports. Then if we put

$$b_n = \max_{2^n \leq k < 2^{n+1}} |a_k|,$$

it is easy to see using the Cauchy criterion that the series (7) converges uniformly and absolutely, provided that

$$\sum_{n=1}^{\infty} b_n 2^{-n/2} < \infty. \tag{8}$$

If the growth condition on $|a_k|$ is satisfied, we have $|b_n| \leq C 2^{\epsilon n}$, and so (8) certainly holds.

For the proof of Theorem 1, we will choose $\{a_k\}$ to be a sequence of independent, standard normal random variables; such a sequence satisfies the condition of Lemma 1 almost surely, with a lot to spare. A more than adequate estimate is very simple to obtain; stronger ones hold but are not necessary here.

Lemma 2. *If $\{Z_k\}$ are standard normal random variables, then*

$$P[|Z_k| = O(k^{\epsilon})] = 1 \quad \text{for any } \epsilon > 0. \tag{9}$$

Proof. A normal distribution has finite moments of every order, so we can use Chebychev's inequality in the form (6.7) to obtain

$$P(|Z_k| \geq k^{\epsilon}) \leq \frac{E(Z_k^{2N})}{k^{2\epsilon N}} = \frac{1 \cdot 3 \cdot 5 \cdots N}{k^{2\epsilon N}}.$$

For any $\epsilon > 0$, the sum on k of the right-hand side converges provided we use a high enough moment N. Thus by the first part of the Borel–Cantelli lemma, with probability 1 only finitely many of the events $\{|Z_k| > k^{\epsilon}\}$ will occur—which implies (9). (Independence of the $\{Z_k\}$ plays no role.)

We return to Theorem 1. The almost sure uniform convergence of (6) follows directly from Lemmas 1 and 2, and since the functions $S_k(t)$ are each continuous and vanish at $t = 0$, the sum $x_t(\omega)$ has the same properties. Thus only conditions (ii) and (iii) of the definition, stating that $\{x_t\}$ has normally distributed, independent increments, remain to be proved.

This verification is quite easy given two facts about multidimensional normal distributions that are stated as propositions in the postscript to this section. (The only other use of these results will come soon in Section 23.) Consider any set $t_1, \ldots, t_n$ in $[0, 1]$ and the corresponding random variables x_{t_j}. If the series in (6) were to be terminated after the Kth term, it would define the random variables x_{t_j} as linear combinations of $Z_0, Z_1, \ldots, Z_K$; those (modified) x_t's would then have a joint normal distribution by definition (see below). The convergent infinite series (6) thus defines the actual x_{t_j} as limits of jointly normal variables, and so they themselves have a joint normal distribution by Proposition 2. According to Proposition 1, this distribution is completely determined by the covariances $E(x_{t_j} x_{t_k})$. If the process $\{x_t\}$ defined by (6) really is Brownian motion, its covariance function should be $E(x_s x_t) = \min(s, t)$ (recall Problem 21.2), and conversely. But the covariance can be calculated quite easily:

$$E(x_s x_t) = E\left(\sum_{j=0}^{\infty} Z_j S_j(s) \sum_{k=0}^{\infty} Z_k S_k(t)\right) = \sum_{j,k} E(Z_j Z_k) S_j(s) S_k(t)$$

$$= \sum_{j=0}^{\infty} S_j(s) S_j(t) = \sum_{j=0}^{\infty} \int_0^s H_j(u)\, du \int_0^t H_j(v)\, dv$$

$$= \sum_{j=0}^{\infty} (\mathbf{1}_{[0,s]}, H_j)(\mathbf{1}_{[0,t]}, H_j) = (\mathbf{1}_{[0,s]}, \mathbf{1}_{[0,t]}) = \min(s, t), \quad (10)$$

where the evaluation of the sum on the last line is simply an application of Parseval's relation. This completes the proof and shows that Brownian motion exists—for the interval $[0, 1]$. □

How about $[0, \infty)$? Notice that the measure-theoretic part of our construction is completely contained in the sentence "Let $Z_0, Z_1, \ldots$ be a sequence of independent, standard normal random variables...." Clearly, by renumbering or otherwise, we can produce a countable number of such sequences, independent of each other. Using these and relation (6), we can then construct a sequence of independent random functions $\{x_t^{(n)}(\omega)\}$, each defined for $0 \le t \le 1$. Finally, we can piece these together to produce a function on $[0, \infty)$. Specifically, define $x_t(\omega) = x_t^{(1)}(\omega)$ for $0 \le t \le 1$, and proceed inductively by setting

$$x_t(\omega) = x_n(\omega) + x_{t-n}^{(n+1)}(\omega) \quad (11)$$

when $t \in [n, n+1)$. This determines a continuous function for all $t \in [0, \infty)$. It should be fairly evident that this function satisfies the conditions of a Brownian motion for all $t \geq 0$, and we will omit the details of the verification. This finishes the first construction of the Brownian motion process.

Postscript: Normal Distributions in R^k

Let $Z_1, \ldots, Z_n$ be independent random variables with the standard normal distribution. (That is, they all have variance 1 and expected value 0.) We define a random point Z in R^n by choosing these variables for its rectangular coordinates. The distribution of that point is called the *standard normal distribution* in n dimensions. Now suppose A is any linear mapping from R^n into some Euclidean space R^k (which may or may not be distinct from R^n); $A(Z)$ then represents a random point in R^k. Let P denote the distribution of this point.

Definition 1. *A probability measure P on R^k is called a* normal distribution *(with expected value 0) if and only if it is the distribution of a random point $A(Z)$ that can be represented as described above. Scalar random variables $Y_1, \ldots, Y_k$ are said to be* jointly normally distributed with mean 0 *if and only if they are the coordinates of such a point $A(Z)$.*

In matrix notation, suppose the coordinates of Z form a column vector and that A is represented by a $k \times n$ matrix. Then $Y = AZ$ will be a random column vector in R^k, again with mean 0. Its covariance matrix is easily found:

$$C = E(YY^T) = E(AZZ^T A^T) = AA^T \tag{12}$$

since the covariance matrix of Z is the identity. The matrix C, like any covariance matrix, is symmetric and nonnegative definite; it will be invertible if and only if $\operatorname{rank}(A) = k$. Conversely, since every nonnegative definite matrix C can be written as AA^T for some $k \times k$ matrix A, it follows that every such C is the covariance matrix of the normal random vector AZ, where Z has the standard normal distribution in R^k. Of course, C may also be the covariance of other normal vectors $A'Z'$, where Z' is a standard normal in some other space R^m and A' is a linear mapping of R^m into R^k. As we will very soon see, however, when this happens $A'Z'$ has the same distribution as AZ.

The facts about normal distributions needed here are only two, and we will prove them both using characteristic functions. (See Postscript 2 to Section 15).

Proposition 1. *A normal distribution on R^k (with mean 0) is uniquely determined by its covariance matrix C.*

Proposition 2. *The family of all normal distributions on R^k with mean 0 is closed under weak convergence.*

Proof of Proposition 1. We suppose that $Y = AZ$ is any normal random vector in R^k and proceed to calculate its characteristic function. It is easy to obtain the characteristic function of Z in R^n; since the components are independent normals, we have

$$\phi(\lambda_1, \ldots, \lambda_n) = E(e^{i \sum \lambda_j Z_j}) = e^{-\frac{1}{2} \sum \lambda_j^2}.$$

Expressed in matrix form, this becomes

$$\phi(\lambda) = E(e^{i\lambda^T Z}) = e^{-\frac{1}{2} \lambda^T \lambda}, \tag{13}$$

where λ is the column vector in R^n with components $\lambda_1, \ldots, \lambda_n$ and the "exponent" T means transpose. But then if ψ denotes the characteristic function of Y and ξ is a column vector in R^k, we have

$$\psi(\xi) = E(e^{i\xi^T Y}) = E(e^{i\xi^T AZ}).$$

This is the same as the first expression for $\phi(\lambda)$ in (13) with $\lambda^T = \xi^T A$; hence

$$\psi(\xi) = e^{-\frac{1}{2} \xi^T AA^T \xi} \tag{14}$$

which depends only on the covariance $C = AA^T$. Since the distribution of Y is in turn determined by its characteristic function (Theorem 15.7), this proves Proposition 1. □

Corollary 1. *Suppose that (scalar) random variables $Y_1, \ldots, Y_k$ are jointly normally distributed. Then if these Y_j are uncorrelated, they are independent.*

Proof of Proposition 2. Suppose that $\{P_n\}$ are normal distributions on R^k with mean 0, and that $P_n \Rightarrow P$. Each P_n has a characteristic function of the form

$$\psi_n(\xi) = e^{-\frac{1}{2} \xi^T C_n \xi},$$

and these functions converge to the characteristic function of P. Taking logs (everything is real), we see that the limit of the quadratic forms $\xi^T C_n \xi$

exists for every ξ; this limit must itself be some nonnegative definite quadratic form $\xi^T C \xi$. (Verify.) This form, multiplied by $-1/2$, is the log of the characteristic function of the limit P, and so the corresponding distribution must be normal.   □

Remark. The multidimensional normal distribution was mentioned once before, in Example 4.3. There it was defined by assuming that the distribution had a probability density function of the form

$$f(x_1, \ldots, x_k) = K e^{-\frac{1}{2} \sum d_{rs} x_r x_s} = K e^{-\frac{1}{2} x^T D x}$$

where $D = [d_{rs}]$ is a symmetric, positive definite matrix and K is a normalizing constant. It is not hard to see that this agrees with the definition above as long as the covariance matrix is nonsingular; it then turns out that $D = C^{-1}$. Since we will make no use of such density functions, we omit the verification.

23. SOME PROPERTIES OF BROWNIAN PATHS

This section will discuss properties that are shared by all the functions $x_t(\omega)$ of a Brownian motion process, except those belonging to some ω set of probability 0. We will prove several things in full and mention a few others without details; many more results of this kind can be found in the interesting book by Paul Lévy (1965). Throughout the section, "Brownian motion" means any stochastic process satisfying the conditions of Definition 21.1.

We begin with the large-scale behavior of the paths. In view of the independent increments property (condition 2 of the definition), it is hardly surprising that the Brownian paths grow in the same way as sums of independent random variables. In particular, they obey the "law of the iterated logarithm."

Theorem 1. *Let $x_t(\omega)$ be a Brownian motion process, $0 \le t < \infty$. Then*

$$P\left(\limsup_{t \to \infty} \frac{x_t(\omega)}{\sqrt{t \log \log t}} = \sqrt{2}\right) = 1. \tag{1}$$

Proof. If we replace the continuous parameter t by an integral one, (1) becomes a special case of the law that was discussed in Section 12. In this situation $S_n = x_n$, which means that we must choose $X_k = x_k - x_{k-1}$; by the definition of Brownian motion, these random variables are independent and have the standard normal distribution. Estimates 3 and 4 of Section 12 then assert that the lim sup in (1) holds as $t \to \infty$ through positive integer values n.[9] Since the lim sup can only increase if t is allowed to grow through *all* positive values, we see that (1) must at least hold with $\geq \sqrt{2}$ in place of $= \sqrt{2}$.

In fact, the lim sup in (1) is the same when t is restricted to integers as when t takes on all values, but this is not immediately obvious since, conceivably, big excursions of the path for values of t between the integers could make the lim sup larger in the continuous case. To rule out this possibility, we will need a familiar sort of estimate.

Lemma 1. *For any $a > 0$,*

$$P\left(\max_{0 \leq t \leq 1} x_t > a\right) \leq 2P(x_1 \geq a). \tag{2}$$

Proof. Let n be any positive integer, and notice that by writing

$$x_{k/2^n} = \sum_{j=1}^{k} [x_{j/2^n} - x_{(j-1)/2^n}]$$

we exhibit $x_{k/2^n}$ as the sum of k independent random variables, each normally distributed with mean 0 and variance 2^{-n}. Thus by (12.15), the improved form of Lemma 12.2 which holds for symmetric distributions such as (in this case) the normal, we have

$$P\left(\max_{k \leq 2^n} x_{k/2^n} > a\right) \leq 2P(x_1 \geq a). \tag{3}$$

But when n increases, the events (sets) on the left side of (3) also increase while the upper bound on the right side is independent of n; thus the same bound holds for the union of those events. That union is the event that $x_t > a$ for some dyadic rational number t between 0 and 1. For all ω such that $x_t(\omega)$ is continuous in t, this is the same as $\max_{t \in (0,1)} x_t > a$. Thus since continuity holds except for a set of probability 0, the bound in (2) follows.

---

[9]Estimate 4 was not fully proved in Section 12. However, in the present (normal) case the missing step (the proof of Lemma 3) is easily supplied, as was indicated in Problem 12.5.

Remark. A slightly weaker version of (2) in which a on the right side is replaced by $a - \sqrt{2}$ follows from Lemma 12.2 as stated (without the improvement indicated in Problem 12.2), and this weaker version would suffice here. But Lemma 1 is preferable as written because (2), in fact, gives not only an upper bound but the exact value for the probability of a Brownian excursion above the level a. This will become clear from the discussion of Section 25.

It is not difficult to finish the proof of Theorem 1. Let

$$m_n = \max_{n-1 \le t \le n} (x_t - x_{n-1}). \tag{4}$$

Then Lemma 1 applies to each of the random variables m_n:

$$P(m_n > a) \le 2P(x_n - x_{n-1} \ge a) = \frac{2}{\sqrt{2\pi}} \int_a^\infty e^{-u^2/2}\, du.$$

From this fact plus Lemma 22.2 of the previous section, we have $m_n = O(n^\epsilon)$ with probability 1, and so in particular

$$P[m_n = o(\sqrt{n \log \log n})] = 1. \tag{5}$$

This is the bound that is needed. We know that the upper bound in (1) holds a.s. as $t \to \infty$ through integer values of t, and (5) shows that no excursions of the path which take place between integer values of t can affect the validity of (1) for all t. This finishes the proof. □

Corollary 1. *The function x_t has (a.s.) arbitrarily large zeros.*

Proof. Theorem 1 shows that with probability 1, x_t is positive for arbitrarily large values of t. The same thing is clearly true for the process $\{-x_t\}$ since it is also a Brownian motion, and so x_t changes sign infinitely often as $t \to \infty$. Since the paths are continuous functions, these sign changes imply that x_t must have zeros for arbitrarily large t as well. □

We turn now to some local properties of the paths, starting with differentiability. The difference quotient $(x_{t+h} - x_t)/h$ has a normal distribution with mean 0 and variance $1/h$; it follows that

$$\lim_{h \to 0} P\left(\left| \frac{x_{t+h} - x_t}{h} \right| \le M \right) = 0 \tag{6}$$

for each M. This shows that $P(dx_t/dt \text{ exists}) = 0$ for each t.

Remark 1. Sometimes, it might be necessary to distinguish between "almost sure" derivatives, "in the mean" derivatives and "in probability" derivatives, depending on how the difference quotient converges to x'_t. In the present case, (6) implies that none of these can exist.

Remark 2. Since x_t is supposed to represent the position at time t of a physical particle, it may seem paradoxical that dx_t/dt, which ought to give the velocity of the particle, does not exist. In reality, this simply means that the mathematical model doesn't describe the physical phenomenon exactly for very small intervals of time. Other stochastic processes have been proposed to give a more accurate representation; in models due to L. S. Ornstein and G. E. Uhlenbeck, the velocity dx_t/dt does exist but the second derivative, representing acceleration, does not. Still, up to a point the extreme irregularity of the paths in the original model *is* physically plausible. Even before the theory had been elaborated, at least one observer commented that the paths of the (physical) Brownian motion "resemble the nondifferentiable functions of the mathematicians."

The finding of nondifferentiability can be strengthened in several ways. One of them is investigating the exact magnitude of the small fluctuations of x_t; the result is known as the "local law of the iterated logarithm."

Theorem 2. *For each $t_0 > 0$,*

$$P\left(\limsup_{h \to 0+} \frac{x_{t_0+h} - x_{t_0}}{\sqrt{h \log \log(1/h)}} = \sqrt{2}\right) = 1. \tag{7}$$

The neatest proof (not the original one) is based on a transformation that relates the local and large-scale behavior of the Brownian paths.

Proposition 1 (Lévy). *Let $\{x_t\}$ be a Brownian motion process and define*

$$y_t = \begin{cases} tx_{1/t} & \text{for } t > 0; \\ 0 & \text{for } t = 0. \end{cases} \tag{8}$$

Then $\{y_t\}$ is also a Brownian motion.

Proof. The continuity of y_t for $t > 0$ is immediate from that of x_t, and continuity at $t = 0$ follows from Theorem 1. Clearly, any finite set of the y_t have a joint normal distribution since that is true of the x_t. Hence we can check properties 2 and 3 of Definition 21.1 by calculating the covariance function of $\{y_t\}$ and appealing to Proposition 1 of the previous section. For $t \leq s$, we find that

$$E(y_t y_s) = ts E(x_{1/t} x_{1/s}) = ts \min(t^{-1}, s^{-1}) = t = \min(t, s),$$

which is again the correlation function of Brownian motion. □

The proof of Theorem 2 (with $t_0 = 0$) is immediate from this proposition and Theorem 1. For the general case, we apply Proposition 1 to the process $x_{t_0+t} - x_{t_0}$, which is also Brownian motion. It is also easy to deduce the following.

Corollary 2. *For each $t_0 \geq 0$, the function $x_t - x_{t_0}$ has (a.s.) a sequence of zeros approaching t_0 from the right.*

In particular, $t = 0$ is not an isolated zero of x_t. It is tempting to conclude that the function x_t has *no* isolated zeros, arguing that we could take any t such that $x_t = 0$ as the t_0 of the corollary. This is not really a valid proof, since the t_0 in Theorem 2 must be a fixed value, not a random variable. The conclusion is nevertheless true, and it can be proved by justifying the intuitive idea that upon reaching state 0 at the time $t_0 > 0$, the process "starts over" from scratch. This is a special case of the so-called *strong Markov property*. The next section introduces the concept of Markov processes, but the strong property is beyond the scope of this book [see e.g., Lamperti (1977), Chapter 9].

A different extension of the nonexistence of x_t' for each t was given by Weiner himself: *Almost all paths x_t are nowhere differentiable in t.* (Even more is true: the functions $x_t + ct$ have no points of increase or decrease for any choice of c.) The proof of nowhere differentiability is intricate, however, and we will settle for a weaker result whose proof is short and neat—except for one technical point that will be clarified afterward.

Theorem 3. *With probability 1, the set of t for which x_t' exists has Lebesgue measure 0.*

Proof. Define the random function

$$f(t, \omega) = \begin{cases} 1 & \text{if } \dfrac{d}{ds} x_s(\omega) \text{ exists at } s = t; \\ 0 & \text{otherwise.} \end{cases} \quad (9)$$

Assuming that f is measurable as a function of (t, ω) with respect to the product σ-field of the Borel sets of R^1 and $\mathcal{B}$ (which we denote by $\mathcal{B}_1 \times \mathcal{B}$), we can apply Fubini's theorem to obtain

$$E\left(\int_0^\infty f(t,\omega)\,dt\right) = \int_0^\infty E[f(t,\omega)]\,dt. \tag{10}$$

But for each t we know that $f(t,\omega) = 0$ a.s. and so $E[f(t,\omega)] = 0$; thus the right-hand side of (10) vanishes and so the left side is 0 as well. But since the random variable $M = \int_0^\infty f(t,\omega)\,dt$ is nonnegative, it follows from $E(M) = 0$ that $M = 0$ a.s.; this is the same as the conclusion of the theorem.

The question of measurability remains; this property holds very generally, not just for Brownian motion. For example, we have the following.

Proposition 2. *Let $\{z_t\}$ be a stochastic process with the property that the path function $z_t(\omega)$ is continuous in t for (almost) all ω. Then the function $z_t(\omega)$ is measurable in (t,ω) with respect to $\mathcal{B}_1 \times \mathcal{B}$.*

Proof. For any random variable X, the step function

$$S(t,\omega) = \begin{cases} X(\omega) & \text{if } a \le t < b; \\ 0 & \text{otherwise} \end{cases} \tag{11}$$

is measurable. To show this, we must check that the inverse image of any open interval (η, ∞) belongs to the product field. But if $\eta \ge 0$, we have

$$S^{-1}(\eta,\infty) = \{(t,\omega) : a \le t < b \text{ and } X(\omega) > \eta\} = [a,b) \times X^{-1}(\eta,\infty);$$

this is the product of a Borel set in R^1 [namely the interval $[a,b)$] with a set in $\mathcal{B}$, so it clearly does belong to $\mathcal{B}_1 \times \mathcal{B}$. If $\eta < 0$, the inverse image is the union of the above product set with another such set in which the complement of $[a,b)$ is paired with Ω; this again belongs to the product field. Hence S is measurable.

Now consider the more complex step function

$$z_t^{(n)}(\omega) = \begin{cases} z_{k/n}(\omega) & \text{if } \dfrac{k}{n} \le t < \dfrac{k+1}{n}; \\ 0 & \text{otherwise,} \end{cases}$$

for $k = 0, 1, \ldots, n^2 - 1$. This is the sum of n^2 functions of the form (11) and so it too is measurable. But because of the continuity of paths, we have

$$P\left(\lim_{n\to\infty} z_t^{(n)}(\omega) = z_t(\omega) \text{ for all } t\right) = 1.$$

Since the almost-everywhere limit of a sequence of measurable functions is measurable, this proves the proposition. □

The rest is easy. Knowing as we now do that $x_t(\omega)$ is measurable, we have from standard results on measurability that

$$\limsup_{h \to 0;\, h \text{ rational}} \frac{x_{t+h}(\omega) - x_t(\omega)}{h} = D^+ x_t(\omega)$$

is a measurable function of (t, ω), and, of course, the same is true for $D^- x_t(\omega)$ defined similarly using lim inf. But the set on which two measurable functions are equal has to be a measurable set. The set of agreement for D^+ and D^- is just the set on which $f(t, \omega) = 1$. This completes the proof that f is measurable, and the proof of the Theorem. □

For a function of bounded variation on R^1, the derivative must exist almost everywhere. Combining this with Theorem 3, we have the following.

Corollary 3. *The paths of a Brownian motion process are nonrectifiable (have infinite length) in any interval of time, with probability 1.*

Problem 1. *Let $\{x_t\}$ be Brownian motion. Prove that $\{t \geq 0 : x_t = a\}$ (i.e., the set of times when the process takes on a particular value) has Lebesgue measure 0 with probability 1, for any a. [Hint: Imitate the proof of Theorem 3.]*

We conclude with two problems giving a little more insight into the variation of the Brownian paths.

Problem 2. *Suppose a function $f(t)$ has a continuous derivative on $[a, b]$, $a < b$, and let $a = t_0 < t_1 < \cdots < t_n = b$ be a partition of the interval. Define*

$$U_n = \sum_{k=1}^{n} [f(t_k) - f(t_{k-1})]^2. \qquad (12)$$

If $n \to \infty$ in such a way that $\max(t_k - t_{k-1}) \to 0$, show that $\lim U_n = 0$. [Note: This is not hard; it isn't strictly relevant, either. It's included here in order to put the next problem into context.]

Problem 3. *Let x_t be a Brownian path function, and form the sum W_n defined using this function in place of $f(t)$ in (12). Take the same sort of limit, that is, $\max(t_k - t_{k-1}) \to 0$ as $n \to \infty$. Prove that $W_n \to (b - a)$ in probability. [Hint: Calculate the mean and variance of W_n.]*

24. MARKOV PROCESSES

Brownian motion is an example of a *Markov process*. Informally, this means that if the "state" x_t of the process is given at any time t, then additional knowledge of the "past" of the process (i.e., events determined by the values of x_s for $s < t$) has no effect on the conditional probabilities of events in the "future" (events determined by x_s for $s > t$). Discrete-time processes satisfying this "Markov property" include sequences of independent random variables as well as the sequences of partial sums and partial maxima of such variables. An elegant theory of discrete-time "Markov chains" is developed in Feller (1968).[10] The modern theory of general Markov processes is deep and extensive, and in this brief introduction we will barely dent the surface.

Suppose that $\{x_t : t \in T\}$ is a stochastic process (a family of real-valued random variables), where T is a subset of R^1. For any $t \in T$, define the three σ-fields:

$$\mathcal{F}_{\leq t} = \mathcal{B}(\{x_s : s \leq t\}); \quad \mathcal{F}_{=t} = \mathcal{B}(x_t); \quad \mathcal{F}_{\geq t} = \mathcal{B}(\{x_s : s \geq t\}). \tag{1}$$

Thus, the fields $\mathcal{F}_{\leq t}$ and $\mathcal{F}_{\geq t}$ consist of all the sets or events defined in terms of the past (or the future) of the process before (after) the moment t, while $\mathcal{F}_{=t}$ consists of those sets that are defined in terms of the state or value of the process at the moment t, that is, the "present." Using these three σ-fields, we can give a rigorous statement of the property described above.

Definition 1. *The stochastic process $\{x_t : t \in T\}$ has the* **Markov** *property (or is a* **Markov** *process), provided that for each $t \in T$ and each set $A \in \mathcal{F}_{\geq t}$,*

$$E(\mathbf{1}_A \mid \mathcal{F}_{\leq t}) = E(\mathbf{1}_A \mid \mathcal{F}_{=t}) \quad \text{a.s.} \tag{2}$$

In words, the conditional probability of any future event A given the entire past and present of the process is the same as the conditional probability of that event given only the present. A process with this property is somewhat analogous to a system in classical mechanics, where knowledge

---

[10] A Markov process is called a "Markov chain" when both the parameter t and the state space (the range of the random variables x_t) are discrete; sometimes the term is also used when the states are discrete but time is continuous.

MARKOV PROCESSES

of the current state (all positions and moments) of the system at some moment is sufficient to calculate its entire future trajectory, and additional knowledge of the past contributes nothing more.

A different, but related, intuitive description of the Markov property is the demand that if we are given the present state of a process, then past and future events must be conditionally independent. In terms of the σ-fields (1), this can be stated as

$$E(\mathbf{1}_A \mathbf{1}_B \mid \mathcal{F}_{=t}) = E(\mathbf{1}_A \mid \mathcal{F}_{=t}) E(\mathbf{1}_B \mid \mathcal{F}_{=t}) \quad \text{(a.s.)} \qquad (3)$$

whenever $A \in \mathcal{F}_{\geq t}$ and $B \in \mathcal{F}_{\leq t}$, for each $t \in T$. We will first prove the following.

Theorem 1. *Conditions (2) and (3) are equivalent.*

Proof.[11] Condition (2) implies (3): Let B be any past event and multiply (2) by $\mathbf{1}_B$. On the left-hand side, the function can be moved inside the conditional expectation since it is measurable with respect to $\mathcal{F}_{\leq t}$; hence,

$$E(\mathbf{1}_A \mathbf{1}_B \mid \mathcal{F}_{\leq t}) = \mathbf{1}_B E(\mathbf{1}_A \mid \mathcal{F}_{=t}).$$

Now take the conditional expectation of both sides with respect to $\mathcal{F}_{=t}$, which is a subfield of $\mathcal{F}_{\leq t}$. The result is

$$E(\mathbf{1}_A \mathbf{1}_B \mid \mathcal{F}_{=t}) = E(\mathbf{1}_B E(\mathbf{1}_A \mid \mathcal{F}_{=t}) \mid \mathcal{F}_{=t}) = E(\mathbf{1}_A \mid \mathcal{F}_{=t}) E(\mathbf{1}_B \mid \mathcal{F}_{=t})$$

since $E(\mathbf{1}_A \mid \mathcal{F}_{=t})$ is measurable with respect to $\mathcal{F}_{=t}$. This asserts the conditional independence of past and future, given the present.

Condition (3) implies (2): To prove (2), we must show that $E(\mathbf{1}_A \mid \mathcal{F}_{=t})$ will serve as a version of $E(\mathbf{1}_A \mid \mathcal{F}_{\leq t})$. The measurability condition is obvious, so it is just necessary to show that

$$\int_B E(\mathbf{1}_A \mid \mathcal{F}_{=t}) \, dP = \int_B \mathbf{1}_A \, dP = P(A \cap B) \qquad (4)$$

---

[11] What follows depends on the properties of conditional expectation established in Theorem 4.2, and we will use these freely without specific references. Conditional expectations, in general, are defined only up to a set of measure 0; when two such expectations are asserted to be "equal" (=), the "almost surely" will usually be taken for granted.

for any past set $B \in \mathcal{F}_{\leq t}$. But using (3) we have

$$P(A \cap B) = \int_\Omega \mathbf{1}_A \mathbf{1}_B \, dP = \int_\Omega E(\mathbf{1}_A \mathbf{1}_B \mid \mathcal{F}_{=t}) \, dP$$

$$= \int_\Omega E(\mathbf{1}_A \mid \mathcal{F}_{=t}) E(\mathbf{1}_B \mid \mathcal{F}_{=t}) \, dP$$

$$= \int_\Omega E[\mathbf{1}_B E(\mathbf{1}_A \mid \mathcal{F}_{=t}) \mid \mathcal{F}_{=t}] \, dP$$

$$= E[\mathbf{1}_B E(\mathbf{1}_A \mid \mathcal{F}_{=t})] = \int_B E(\mathbf{1}_A \mid \mathcal{F}_{=t}) \, dP.$$

This establishes (4), and hence (2). □

Remark. Condition (3) is symmetric with respect to past and future, whereas Definition 1 is not. This implies that (2) has another form in which the roles of past and future are interchanged. Another consequence is that the Markov property is preserved if the direction of time is reversed; in other words, the process $\{x_{-t} : t \in -T\}$ is a Markov process whenever $\{x_t\}$ is one.

Some questions now arise. Of course, we would like to show that Brownian motion really is a Markov process, as has been supposed. Three methods for constructing Brownian motion were promised in Section 21, and only one has been revealed so far. It is also natural to wonder how other interesting Markov processes come about. All these matters can be approached through the concept of a *transition function*, analogous to the transition probability matrix of a discrete Markov chain. The idea is that a function $p_t(x, E)$ should represent the probability of a "transition" by an underlying stochastic process from "state" x into the set of states E after a lapse of time t. Since we are not seeking great generality, T will be $[0, \infty)$; the "states" (i.e., the values of the random variables x_t) are still real numbers.

Definition 2. *A function $p_t(x, E)$, where $t \geq 0$, $x \in R^1$, and E is a Borel subset of R^1, is a stationary Markov transition function provided it satisfies:*

1. *For each t and E, $p_t(x, E)$ is a Borel-measurable function of x.*
2. *For each t and x, $p_t(x, \cdot)$ is a probability measure on R^1.*

MARKOV PROCESSES

3. $p_0(x, \cdot)$ is the measure placing unit mass at the point x.[12]
4. For each $s, t > 0$ and each x and E, p_t satisfies

$$p_{t+s}(x, E) = \int_{R^1} p_t(x, dy) p_s(y, E). \tag{5}$$

Condition (5) is often called the "Chapman–Kolmogorov equation." The motivation behind it is that a transition of the (assumed) underlying process from x into E in time $t + s$ can be broken up into first a transition from x into some intermediate state y in time t, followed by the transition from y into E in the remaining time s; the integral represents the "sum" over all possible choices for y. But so far this is *only* motivation, since we do not yet have a stochastic process.

Example 1. Define $p_t(x, E)$ by means of a normal probability density:

$$p_t(x, E) = \int_E \frac{1}{\sqrt{2\pi t}} e^{-(y-x)^2/2t} \, dy. \tag{6}$$

This is the transition function of the Brownian motion process.

Problem 1. Verify that (6) defines a Markov transition function.

Example 2. Suppose that the "state space" is the integers, that is, substitute Z for R^1 in Definition 2. Then $p_t(x, E)$ is replaced by $p_t(i, j)$, where for every $t \geq 0$, $[p_t(i, j)]$ forms a stochastic matrix $\mathbf{P_t}$.[13] Condition (5) becomes

$$p_{t+s}(i, j) = \sum_k p_t(i, k) p_s(k, j) \quad \text{or} \quad \mathbf{P_{t+s}} = \mathbf{P_t P_s}. \tag{7}$$

One concrete example is the transition function for the *Poisson process*:

$$p_t(i, j) = e^{-ct} \frac{(ct)^{j-i}}{(j-i)!} \quad \text{for } j \geq i \tag{8}$$

[and $p_t(i, j) = 0$ for $j < i$], where c is a positive constant.

---

[12] Continuity in t is often assumed as well; the usual requirement is that for each x the measures $p_t(x, \cdot)$ should converge weakly to $p_0(x, \cdot)$ as $t \to 0+$.
[13] A square matrix $\mathbf{P}$ is *stochastic* or *Markov* if its entries are nonnegative and the sum of each of its rows is 1.

Problem 2. *Show that the matrices defined in* (8) *form a Markov transition function on* Z.

More generally, let $\mathbf{P} = [p(i, j)]$ be any stochastic matrix, let $\mathbf{I}$ denote the identity, and define the matrix $\mathbf{P}_t$ to be

$$\mathbf{P}_t = e^{-ct} e^{ct\mathbf{P}} = e^{-ct} \sum_{n=0}^{\infty} \frac{(ct)^n \mathbf{P}^n}{n!}. \quad (9)$$

Then $\mathbf{P}_t$ is a transition function that describes a Markov chain in continuous time (but not the most general one if the number of states is infinite). This chain can be thought of as a discrete-time Markov chain whose transitions take place at random times governed by a "Poisson clock." If the matrix $\mathbf{P}$ is specified by setting $p(i, j) = 1$ when $j = i + 1$ (and 0 otherwise), then (9) reduces to (8). This construction works for infinite matrices $\mathbf{P}$ but the reader is (only) asked to justify it for finite ones.

Problem 3. *Show that the series in* (9) *defines a transition function for any finite stochastic matrix* $\mathbf{P}$. *[Hint: The hardest part is showing that $p_t(i, j) \geq 0$. Show this first for small values of t, assuming that $\mathbf{P}$ is strictly positive. The condition $p_t(i, j) \geq 0$ then follows for all t. (Why?) Finally, remove the assumption of strict positivity.]*

The construction of a process from a Markov transition function follows the path outlined in Section 5. The transition function, together with a choice of some x_0 to serve as the starting state,[14] is used to define the family of finite-dimensional distributions, and then Kolmogorov's theorem guarantees the existence of the probability space and random variables that form an actual stochastic process. The one-dimensional distributions are chosen to be $P_t(E) = p_t(x_0, E)$, where E is any Borel set in R^1 (in particular, the distribution for $t = 0$ is a unit mass at x_0). To define the joint distributions, let $0 < t_1 < \cdots < t_n$ be any finite set of parameter values, and let S be the "rectangle" in R^n obtained by restricting the kth coordinate to the set E_k, for $k = 1$ to n. Then define

$$P_{t_1, t_2, \ldots, t_n}(S) = \quad (10)$$

$$\int_{E_{n-1}} \cdots \int_{E_2} \int_{E_1} p_{t_1}(x_0, dy_1) p_{t_2 - t_1}(y_1, dy_2) \cdots p_{t_n - t_{n-1}}(y_{n-1}, E_n).$$

As long as these distributions are consistent, Theorem 5.1 (Kolmogorov's theorem) will apply and the process $\{x_t\}$ must exist.

---

[14] Alternatively, a distribution can be specified so that the initial state will be random.

MARKOV PROCESSES 155

Why does the consistency hold? We will illustrate with a special case. Let E be a set of reals, and let τ be the projection from R^2 into R^1 such that $\tau(x, y) = y$; then

$$\tau^{-1}(E) = \{(x, y) \in R^2 : y \in E\}.$$

We need to show that

$$P_{s,t}(\tau^{-1}(E)) = P_t(E) \qquad \text{for } s < t; \tag{11}$$

this is a particular case of the consistency of the finite-dimensional distributions. But (11) follows at once from the Chapman–Kolmogorov equation (5):

$$P_{s,t}(\tau^{-1}(E)) = \int_{R^1} p_s(x_0, dx) p_{t-s}(x, E) = p_t(x_0, E) = P_t(E).$$

Similarly, it is not hard to prove the higher-dimensional analogs of (11), again using (5); these are the consistency conditions needed for Theorem 5.1. We omit the details of this verification.

Theorem 2. *The process $\{x_t\}$ constructed above is a Markov process in the sense of Definition 1 and satisfies*

$$P(x_{t+s} \in E \mid \mathcal{F}_{\leq t}) = p_s(x_t(\omega), E) \quad \text{(a.s.)} \tag{12}$$

for any Borel set E in R^1.

Sketch of Proof. We must verify condition (2) above for a general future set A, but we start with one of the special form

$$A = \{\omega : x_{t+s} \in E\}, \qquad t, s > 0. \tag{13}$$

We will show that

$$P(A \mid \mathcal{F}_{=t}) = p_s(x_t(\omega), E) = P(A \mid \mathcal{F}_{\leq t}), \tag{14}$$

where, of course, each equality is to hold almost surely. We begin with the second equality in (14), and the first part comes along automatically.

It is obvious that $p_s(x_t(\omega), E)$ is measurable with respect to $\mathcal{F}_{=t}$, and hence with respect to the larger field $\mathcal{F}_{\leq t}$, since $p_s(u, E)$ is a measurable function of u. What must be verified, therefore, is that for any past set

$B \in \mathcal{F}_{\leq t}$,

$$\int_B p_s(x_t(\omega), E) \, dP = P(A \cap B). \tag{15}$$

This, in turn, will be proved in stages. Assume first that B is of the form

$$B = \{\omega : x_{t_1} \in E_1, x_{t_2} \in E_2, \ldots, x_{t_n} \in E_n\}, \tag{16}$$

where $0 < t_1 < t_2 < \cdots < t_n = t$ and the $\{E_k\}$ are Borel sets of reals. For this set B, the left and right sides of (15) can be written out explicitly using the definition (10). When this is done, it is clear that (15) holds.

But now notice that both sides of (15) are actually measures when considered as functions of the set B with everything else fixed. These measures are equal for all sets of the form (16), and of course also for all finite disjoint unions of such sets; these sets form a Boolean algebra. By the uniqueness part of Theorem 1.1 (the basic extension theorem), therefore, the two sides of (15) must agree on the σ-field generated by these sets; that field is exactly $\mathcal{F}_{\leq t}$. This verifies the last part of (14). The first part is immediate, since as noted above $p_s(x_t(\omega), E)$ is measurable with respect to $\mathcal{F}_{=t}$.

The next step is to verify (2) for more complicated future sets A that depend on a finite number of future random variables instead of only one. This can be done by induction on the number of such variables, and we omit the details. Finally, the extension to general future sets is made using again the uniqueness part of Theorem 1.1. The details of these arguments can be supplied by the reader or found in Lamperti (1977), Chapter 8. □

Remark 1. If the transition function is the one defined in (6), then the process $\{x_t\}$ has the same finite-dimensional distributions as Brownian motion. As shown in Section 5, however, the σ-field $\mathcal{B}$ used in this construction is too small for many purposes, and in particular it does not contain the set of all continuous functions. It is therefore not sensible to ask whether the process associated with a particular transition function *has* continuous paths. Rather, we should ask whether or not *some* process with the same finite-dimensional distributions has continuous paths; that is, whether it is possible to extend the measure determined on $\mathcal{B}$ by the given transition function to a larger σ-field in such a way that the set of continuous functions becomes measurable and has measure 1. As we have seen, for Brownian motion the answer is "yes." It is natural to seek criteria on the function $p_t(x, E)$ that guarantee such a possibility, and this problem has of course been studied. Even if the criterion for continuity does not hold, under milder restrictions on $p_t(x, E)$ there will be a process whose paths are at least right-continuous with probability 1. [See Lamperti (1977), Chapters 8–10, for a fuller discussion of these questions.] A Markov process with (a.s.) continuous paths is often called a *diffusion* process.

Remark 2. Where do transition functions come from? In discrete time they can be generated by choosing any function $p(x, E)$ to represent the probability of a transition from state x into the set E in one step; this function need only be measurable in x for each E and a probability measure as a function of the set E for each x. The probabilities for n steps are then obtained by iterating this function in a manner similar to matrix multiplication. For example,

$$p^{(2)}(x, E) = \int_{R^1} p(x, dy) \, p(y, E)$$

is the probability of moving from x into E in two steps. This construction generalizes the Markov matrix and its powers that give the transition probabilities for a discrete Markov chain.

When the time parameter is continuous, matters are more complicated. The difficulty is that transition probabilities cannot be chosen arbitrarily for any $t > 0$ since they must always satisfy condition (5). In the case of diffusion processes, the transition probabilities are solutions of certain partial differential equations; for example, the Brownian motion transition function (6) satisfies the heat equation. More generally, transition functions are associated with semigroups of operators on function spaces, and the theory of semigroups and their generators is key to an understanding of Markov processes. An introduction to this theory can be found in Lamperti (1977), Chapters 6 and 7.

25. BROWNIAN MOTION AND LIMIT THEOREMS

In these last sections, we return to the particular case of Brownian motion in order to sketch a few more recent developments. Proofs will often be incomplete or omitted entirely as we briefly survey some results that provide interesting connections with other areas of mathematics.

Suppose that $\{S_k\}$ is the simple random walk mentioned in Sections 17 and 20; that is, $S_k = X_1 + \cdots + X_k$ is the sum of k independent random variables taking the values $+1$ or -1 with probabilities $1/2$. It was recognized long ago that if the walk is speeded up so that many steps take place in unit time while the size of the steps tends appropriately to 0, the walk in some sense tends toward Brownian motion as a limit. This idea was described and exploited by L. Bachelier early in the twentieth century.

But what exactly does "tends toward" mean? For any positive integer n, define the random step function $x_t^{(n)}$ by setting

$$x_t^{(n)}(\omega) = \frac{S_k}{\sqrt{n}} \quad \text{for} \quad \frac{k}{n} \leq t \leq \frac{k+1}{n}, \quad k = 0, 1, 2, \ldots. \quad (1)$$

The de Moivre limit theorem implies that as $n \to \infty$, the distribution of the random variable $x_t^{(n)}$ for any $t > 0$ converges to a normal distribution with mean 0 and variance t; that is, to the distribution of $x_t(\omega)$, where $\{x_t\}$ is a Brownian motion process. Going a bit further, it is not hard to see that for any finite set of values $t_1 < \cdots < t_k$, the joint distribution of $x_{t_1}^{(n)}, \ldots, x_{t_k}^{(n)}$ converges weakly to the joint distribution of $x_{t_1}, \ldots, x_{t_k}$ as defined in Section 21. We therefore say that the scaled down and speeded up random walks $\{x_t^{(n)}\}$ *converge in distribution* to the Brownian motion $\{x_t\}$—that is, all the finite-dimensional distributions of $\{x_t^{(n)}\}$ converge weakly to those of $\{x_t\}$.

Problem 1. *Prove that the joint distribution of $x_s^{(n)}$ and $x_t^{(n)}$ converges weakly to the joint distribution of x_s and x_t, for every positive $s < t$. [Hint: Consider $x_s^{(n)}$ and the difference $x_t^{(n)} - x_s^{(n)}$.] Extend the result to any set $t_1 < \cdots < t_k$.*

The convergence of $\{x_t^{(n)}\}$ to $\{x_t\}$ in distribution suggests that other quantities associated with the path of $x_t^{(n)}$ should be close to the corresponding quantities for the Brownian path x_t; this approximation might provide useful information. For example, consider the largest positive excursion of the random walk during its first n steps, namely $M_n = \max[0, S_1, \ldots, S_n]$. Clearly,

$$\frac{1}{\sqrt{n}} M_n = \max_{0 \leq t \leq 1} \left[x_t^{(n)} \right].$$

In view of Problem 1, it seems plausible that

$$\lim_{n \to \infty} P\left(\max_{0 \leq t \leq 1} [x_t^{(n)}] \leq z \right) = P\left(\max_{0 \leq t \leq 1} [x_t] \leq z \right). \quad (2)$$

If (2) is correct, the limit can be evaluated working either with the Brownian motion process itself or with the random walk $\{S_n\}$.

We can calculate the limiting distribution in (2) explicitly for the simple random walk. The key is the following relation, often called the *reflection principle*:

$$P(\max[0, S_1, \ldots, S_n] \geq m) = 2P(S_n > m) + P(S_n = m) \qquad (3)$$

for any integer $m > 0$. To prove (3), notice that

$$P(M_n \geq m) = P(M_n \geq m;\ S_n > m) + P(M_n \geq m;\ S_n < m)$$
$$+ P(M_n \geq m;\ S_n = m)$$
$$= A + B + C,$$

say. Obviously $A = P(S_n > m)$ and $C = P(S_n = m)$. We also have $A = B$. The reason is that every sequence $S_1, \ldots, S_n$ such that $S_n > m$ has a unique counterpart (with the same probability 2^{-n}) obtained by "reflecting" the path around the line $y = m$ after the first moment when $S_k = m$, so that the reflected path still has a maximum at least m but finishes up below m (see Figure 1).

Combining these facts verifies (3). From this, plus the limiting distribution for S_n given by de Moivre's theorem, it is easy to obtain

$$\lim_{n \to \infty} P(\max[0, S_1, \ldots, S_n] \leq z\sqrt{n}) = N_1(z), \qquad (4)$$

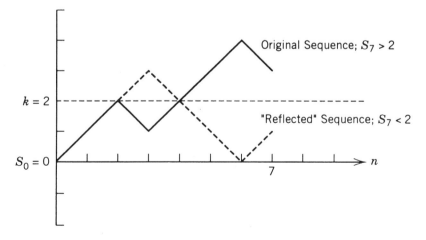

Figure 1. The reflection principle.

where N_1 is the *truncated normal* distribution function

$$N_1(z) = \begin{cases} \dfrac{2}{\sqrt{2\pi}} \int_0^z \exp(-u^2/2)\, du & \text{for } z > 0; \\ 0 & \text{for } z \leq 0. \end{cases} \qquad (5)$$

Then provided (2) holds, N_1 must also be the distribution of the maximum attained by the Brownian motion path during the interval $0 \leq t \leq 1$.

Remark. The reflection argument leading to (3) works because the paths of a random walk are symmetric and "continuous," in the sense that the level m cannot be jumped over. It is plausible to apply these ideas directly to the Brownian motion, and since $P(x_1 = z) = 0$, the result ought to be

$$P\left(\max_{0 \leq t \leq 1}[x_t] \geq z\right) = 2P(x_1 \geq z).$$

[Recall (23.2) and the subsequent remark.] The result is correct but the argument is not rigorous, since the time of first reaching the level z is a random variable, and "reflecting" at that moment requires an appeal to the strong Markov property [see Lamperti (1977), Chapter 9].

It is easy to see that a more general sort of random walk $\{S_m\}$, where $S_m = X_1 + \cdots + X_m$ and $\{X_k\}$ is any sequence of independent, identically distributed random variables with mean 0 and variance 1, also converges in distribution to Brownian motion when it is speeded up and rescaled. This follows just as in Problem 1 if we replace the de Moivre theorem with the general central limit theorem of Section 16. The reader can think through what is involved.

It does not follow automatically, but surely seems plausible, that (2) still holds in this more general situation. If that is true, since we have already identified the distribution of the maximum of the Brownian path, we would obtain the following.

Theorem 1. *Let $\{X_k\}$ be any sequence of independent, identically distributed random variables with mean 0 and variance 1. Then the partial sums $S_k = X_1 + \cdots + X_k$ satisfy*

$$\lim_{n \to \infty} P(\max[0, S_1, \ldots, S_n] \leq z\sqrt{n}) = N_1(z). \qquad (6)$$

This result was one of four limit theorems proved in a 1946 paper by Paul Erdős and Mark Kac.[15] Their "new method of proof"[16] used what the authors called an *invariance principle*. The idea was to show that if a limit relation such as (6) holds for *some* sequence of random variables $\{X_k\}$ satisfying the assumptions, then it must hold for *any* such sequence. Given this, it is only necessary to derive (6) in some particular case (as we have done above) in order to establish it in general. Erdős and Kac derived three more limit theorems using this method, and in another paper published the following year they added a fifth one. The 'invariance principle' had to be proved separately for each result.

This idea of Erdős and Kac was extended by M. Donsker a few years later.[17] Donsker was able to prove an invariance principle similar to (2) but with a *general* functional of the paths $x_t^{(n)}$ and x_t in place of specific functionals such as the maximum. This result, in turn, was put in its proper context a few years later by Yu. V. Prokhorov.[18] One way to state Prokhorov's version is as follows: First, we restrict our attention to times between 0 and 1. (Any finite interval would do as well.) Next, we modify slightly the construction of $x_t^{(n)}$ in (1) in order to make these functions continuous, keeping the same values as before when $t = k/n$ but defining $x_t^{(n)}$ by linear interpolation between these points for other values of t instead of using a step function. This whole function will, of course, be determined by the values of the random variables $X_1, \ldots, X_n$.

Now let C denote the space of continuous functions on $[0, 1]$ with the uniform metric

$$d(x, y) = \max_{0 \le t \le 1} |x(t) - y(t)|.$$

It is evident that the random functions $\{x_t^{(n)}\}$ map the sample space of the $\{X_k\}$ into C, and so they induce probability measures P_n on the Borel sets of the latter space.[19] The Brownian motion $\{x_t\}$ also induces such a

---

[15]"On certain limit theorems in the theory of probability," *Bull. Amer. Math. Soc.* 52 (1946) pp. 292–302.

[16]The method may have been new but this result was not; in particular, a 1938 paper by Kolmogorov foreshadowed many of the results described in this section.

[17]"An invariance principle for certain probability limit theorems," *Memoirs Amer. Math. Soc.* 6 (1951.)

[18]"Convergence of random processes and limit theorems in probability theory," *Theory Prob. Applications* 1 (1956), pp. 157–214.

[19]The "Borel sets" of C (as of any metric space) are, once again, the elements of the smallest σ-field of sets that contains all the open sets. Of course, it needs to be verified that the mappings $\{x_t^{(n)}\}$ from the probability space Ω into C are measurable, but we will omit this detail.

measure, which we call W (for Wiener measure). Then Donsker's result, as reformulated by Prokhorov, amounts to the following.

Theorem 2. *As $n \to \infty$, $P_n \Rightarrow W$ on C.*

The proof of this theorem uses tools developed in this book, but it is rather lengthy and will be omitted. [A complete discussion can be found in Billingsley (1968).] To see some of what the theorem implies, we note an easy fact about weak convergence.

Proposition 1. *Suppose that V and V' are metric spaces, that f is a continuous mapping from V into V', and that P_n and P are probability measures on V such that $P_n \Rightarrow P$. Let Q_n and Q be the corresponding measures induced by f on V'; that is, let*

$$Q_n(E) = P_n(f^{-1}(E)) \quad \text{and} \quad Q(E) = P(f^{-1}(E)),$$

for any Borel set E in V'. Then $Q_n \Rightarrow Q$.

Proof. Let g be any bounded, continuous, real-valued function on V'; to prove the weak convergence, we need to show that

$$\lim_{n \to \infty} \int_{V'} g \, dQ_n = \int_{V'} g \, dQ. \tag{7}$$

But the composition $g \circ f$ is a bounded, continuous function from V to the reals, so since $P_n \Rightarrow P$, we have

$$\lim_{n \to \infty} \int_V g \circ f \, dP_n = \int_V g \circ f \, dP. \tag{8}$$

By Theorem 1 from Section 2.1, however, the integrals in (7) are the same as the corresponding integrals in (8). □

It is not surprising that Theorem 1 is a corollary of Theorem 2; we need only observe that the functional

$$f(x_t) = \max_{0 \le t \le 1} [x_t]$$

is a continuous mapping from C into R^1, and apply the above proposition. Then the measures Q_n will be the distributions of the random variables $f(x_t^{(n)})$, and these distributions must converge weakly to Q, the distribution of $f(x_t)$. Erdös and Kac also considered the functionals

$$\max_{0 \le t \le 1}[|x_t|], \qquad \int_0^1 x_t^2 \, dt, \qquad \text{and} \qquad \int_0^1 |x_t| \, dt;$$

since these are continuous on C with respect to the uniform metric, each of them can be used instead of the maximum to produce a limit theorem similar to Theorem 1. Of course, the limiting distribution will be different in each case; it is the distribution of the particular functional applied to a Brownian path.

Remark 1. The conclusion that $Q_n \Rightarrow Q$ is still valid if the function f is only assumed to be continuous at all points of V except for a set S with $P(S) = 0$, instead of continuous everywhere. This change enlarges the class of limit theorems that follow from Theorem 2. We omit the details [see Billingsley (1968)].

Remark 2. In Section 21 three methods for constructing a Brownian motion process were promised but only two have been mentioned so far. A third method can be based on the approximation by simple random walk. As noted, the rescaled walks $\{x_t^{(n)}\}$ induce measures $\{P_n\}$ on C. It can be shown that this family of measures is conditionally compact, so that it must have at least one limit point—that is, a limiting measure—in the topology of weak convergence. (The proof uses the theorem of Prokhorov stated at the end of Section 13.) But the finite-dimensional distributions of the $\{x_t^{(n)}\}$ converge to those of Brownian motion. This implies that the limiting measure must coincide with the Wiener measure.

Remark 3. The concept of the so-called "invariance principle," properly understood as the weak convergence of the measures induced by a sequence of random processes on some function space, has been applied to limit theorems involving many other processes beside Brownian motion. For example, sequences of random walks that are not spatially homogeneous may converge in distribution to diffusion processes other than Weiner's [see Feller (1968), Chapter XIV], and when this occurs it is natural to try and prove the weak convergence of the corresponding measures on C as well. The concept can also be applied to processes that do not have continuous path functions. In such cases the space C is not appropriate, and it is necessary to consider measures on spaces of discontinuous functions. We will not pursue this thread further; see once again Billingsley (1968).

In the remainder of this section, we describe very briefly a different sort of generalization. As noted above, the convergence of rescaled random walks to Brownian motion is just one among an enormous variety of approximation and limit theorems for stochastic processes. However, the random walk to Brownian motion approximation has a special property, in that a *single* process (the random walk) is *speeded up and rescaled* to produce the processes $\{x_t^{(n)}\}$ that converge to a limit (Brownian motion). One can ask whether, starting with something else in place of simple random walk and perhaps using different rescaling constants, other limiting processes might be obtained.

The answer, naturally, is "yes" and many examples are known. These possible limiting processes occupy a position somewhat analogous to that of the stable distributions. It turns out, in fact, that if such a limiting process is assumed to have stationary independent increments, then the distribution of those increments must *be* stable. Like the stable distributions, these asymptotic processes are characterized by a functional identity. In the case of Brownian motion, for example, it is easy to see that the processes $\{x_t\}$ and $\{x_{at}/\sqrt{a}\}$ have the same finite-dimensional distributions for every $a > 0$. In general, it turns out that a stochastic process $\{y_t\}$ can appear as the asymptote of another process that is being infinitely accelerated and rescaled if and only if it satisfies

$$\{y_t\} \equiv \left\{\frac{y_{at}}{A(a)}\right\} \qquad \text{for every } a > 0, \tag{9}$$

where $A(a) > 0$ is a scaling function and the symbol $\equiv$ means that the two processes have the same finite-dimensional distributions. It then follows that $A(a) = a^p$ for some $p > 0$. Because of the analogy between the role of stable distributions as the possible limiting distributions for partial sums, and that of the processes satisfying (9) as asymptotes, the author once proposed the name *semi-stable* for the latter.[20] Fortunately or not, the doubtless more felicitous term *self-similar* has become the one more commonly used; there is a substantial literature under both names. The structure of stochastic processes satisfying (9) that are also of the Markov type has been analyzed rather completely, at least for one dimension.[21]

---

[20] "Semi-stable stochastic processes," *Trans. Amer. Math. Soc.* 104 (1962), pp. 62–78.
[21] J. Lamperti, "Semi-stable Markov processes," *Z. Wahrscheinlichkeitstheorie verw. Geb.* 22 (1972), pp. 205–225.

Example. [22] Suppose that $\{X_k\}$ are independent, identically distributed random variables, and let $M_n = \max(0, X_1, \ldots, X_n)$. Let us assume that the sequence of maxima has a limiting distribution as in (18.6):

$$F_n(x) = P\left(\frac{M_n - b_n}{a_n} \leq x\right) \Rightarrow F(x), \qquad (10)$$

where F is one of the distributions Φ_α given in (14.5). Define a random step function as in (1) by setting

$$m_t^{(n)} = \frac{M_k - b_n}{a_n} \quad \text{for} \quad \frac{k}{n} \leq t < \frac{k+1}{n}, \qquad k = 1, 2, \ldots; \qquad (11)$$

we also define $m_t^{(n)} = 0$ for $0 \leq t < \frac{1}{n}$. Then it is not hard to show that the processes $\{m_t^{(n)}\}$ converge in distribution as $n \to \infty$ to a limiting process $\{m_t\}$ which satisfies

$$P(m_t \leq x) = F(x)^t \qquad \text{for } t \geq 0 \qquad (12)$$

and

$$p_t(x, (-\infty, y]) = P(m_{t+s} \leq y \mid m_s = x) = \begin{cases} 0 & \text{for } y < x; \\ F(y)^t & \text{for } y \geq x. \end{cases} \qquad (13)$$

This limiting process is Markov, and (13) defines its transition probability function. From (12) and (13) plus the formula (14.5) for F, it is easy to see that the finite-dimensional distributions of $\{m_t\}$ and of $\{m_{at}/a^{1/\alpha}\}$ are identical. Thus $\{m_t\}$ has the self-similar or semi-stable property (9)—as it must due to its role as a limiting process. Incidentally, (12) and (13) also hold if the function F in (10) is one of the other two kinds of maximal distributions, but in these cases the random variable m_0 must be taken as $-\infty$. These limits were named *extremal processes* by M. Dwass.

Problem 2. *Show that (13) defines a Markov transition function for any choice of the distribution function F, and verify that (9) holds with $A(a) = a^{1/\alpha}$ when $F = \Phi_\alpha$.*

26. BROWNIAN MOTION AND CLASSICAL ANALYSIS

There are deep connections between the theory of Markov processes and parts of older, nonprobabilistic analysis. Brownian motion itself is inti-

---

[22] See M. Dwass, "Extremal processes" and J. Lamperti, "On extreme order statistics," *Ann. Math. Stat.* 35 (1964), pp. 1718–1737.

mately related to classical potential theory and other aspects of the Laplace operator. We will not develop these connections systematically here, but just take a look at two striking ways in which Brownian motion can provide new insights into old problems. At times the arguments will be less than rigorous, although they can be fully justified once the "strong Markov property" has been firmly established. Background on the "classical" (non-probabilistic) approach to these problems can be found in Courant and Hilbert (1953) and Rudin (1987); the theory of Markov processes is continued in Lamperti (1977). An extensive discussion of the connection to potential theory was given by Port and Stone (1978).

To begin, we will discuss the *Dirichlet problem* in the following form: Let S be a bounded, connected open set in R^k, and denote its boundary by ∂S. Suppose we are given a continuous function f defined on ∂S. The problem is to find, if possible, a function ϕ that is continuous on $S \cup \partial S$, equals f on ∂S, and that is *harmonic*[23] on S. Solutions can be found in several ways provided the boundary of S is sufficiently smooth, but there are cases where no solution exists for the problem as we have defined it. Norbert Weiner defined a "generalized solution" that always exists; it is a function that is harmonic in S but that takes on the boundary values f in a weaker sense than that implied by the above definition. Weiner did not use the "Weiner process"—that is, Brownian motion—to define his solution, but that now turns out to be the neatest way to do it.[24]

Let $\{\mathbf{x}_t\}$ be a Brownian motion in R^k; this means that each coordinate of the vector $\mathbf{x}_t$ is an ordinary (one-dimensional) Brownian motion and these coordinate processes are independent. Then $\{\mathbf{x}_t\}$ is a Markov process with the transition probability function

$$p_t(\mathbf{x}, \mathbf{E}) = \frac{1}{(2\pi t)^{k/2}} \int_E e^{-|\mathbf{x}-\mathbf{y}|^2/2t} \, dy_1 \cdots dy_k, \qquad (1)$$

where $\mathbf{x}$ and $\mathbf{E}$ are a point and a Borel set of R^k, and $|\mathbf{x} - \mathbf{y}|^2$ denotes the squared distance $\sum (x_j - y_j)^2$ from $\mathbf{x}$ to $\mathbf{y}$. The process can be started at any point of R^k by adding a constant vector to the Brownian motion beginning at $\mathbf{0}$; let us choose a point $\mathbf{u} \in S$ as the initial state. Define the (random)

---

[23] That is, ϕ is twice continuously differentiable and satisfies $\Delta \phi = 0$, where Δ denotes the Laplace operator $\sum_{j=1}^{k} \partial^2 / \partial x_j^2$.

[24] The germ of this idea goes back a long way; it was pointed out by Courant, Friedrichs, and Lewy in 1928 that random walks could be used to solve the discrete analogue of the Dirichlet problem.

exit time as

$$T = \inf\{t > 0 : \mathbf{x}_t \notin S\}; \tag{2}$$

because of the continuity of the paths we have $\mathbf{x}_T \in \partial S$ with probability 1, so that the random variable $f(\mathbf{x}_T)$ is well defined. Now let the function $\phi(\mathbf{u})$ be the expected value of this quantity; that is, define

$$\phi(\mathbf{u}) = E(f(\mathbf{x}_T) \mid \mathbf{x}_0 = \mathbf{u}). \tag{3}$$

Then $\phi(\mathbf{u})$, defined for all $\mathbf{u} \in S$, is the generalized solution to the Dirichlet problem.

What needs to be shown to verify that claim? The expectation does exist since f is bounded,[25] but why is ϕ harmonic? We will sketch the proof using an idea due to S. Kakutani. Let B be the surface of a sphere with its center at $\mathbf{u}$ and contained entirely within S. Let $\tau(B)$ be the instant at which the Brownian motion first reaches B; obviously $\tau(B) < T$ because of path continuity. Since the transition function (1) is invariant under any rotation of axes, it is intuitively clear that the point where the Brownian motion hits the surface B, namely $\mathbf{x}_{\tau(B)}$, will be uniformly distributed over B.

We now appeal to the strong Markov property mentioned earlier: If we are given that the process $\{\mathbf{x}_t\}$ has reached the spherical surface B at a point $\mathbf{v}$, then the conditional expectation of $f(\mathbf{x}_T)$ will be the same as if the process had simply originated at $\mathbf{v}$; in other words, it will be $\phi(\mathbf{v})$. We can thus compute the expectation $\phi(\mathbf{u})$ by *conditioning* on the random point $\mathbf{v} = \mathbf{x}_{\tau(B)}$ which, as we have seen, is distributed uniformly over the surface of B. The conclusion is that the function $\phi(\mathbf{u})$ is equal to the *average* of its values over any small enough spherical surface centered at $\mathbf{u}$. This averaging property implies that ϕ is a harmonic function on S [see Rudin (1987), Chapter 11]. The argument just given can be made fully rigorous once the strong Markov property has been defined and established—but as noted we must leave that story for another day and another book.

We turn to the boundary behavior of ϕ. This is the hardest part in "classical" studies, and the probabilistic interpretation is a great help in obtaining an intuitive understanding of what goes on. The following fact is easy but useful.

Proposition 1. *Let* $\mathbf{r}$ *be a boundary point of* S. *In order that* $\phi(\mathbf{u}) \to f(\mathbf{r})$ *as* $\mathbf{u} \to \mathbf{r}$ *from within* S, *it is sufficient that for any neighborhood* N

---

[25] There is measurability to be checked, but we will take that for granted.

of **r** *we have*

$$\lim_{\mathbf{u}\to\mathbf{r}} P(\mathbf{x}_T \in N \cap \partial S \mid \mathbf{x}_0 = \mathbf{u}) = 1. \tag{4}$$

To verify this, assume (4) holds for some **r**. Choose N small enough so that (by continuity) f is nearly equal to $f(\mathbf{r})$ throughout $N \cap \partial S$. By (4), if the starting point **u** is close enough to **r**, the Brownian motion will hit ∂S at a point belonging to N with high probability; in these cases, $f(\mathbf{x}_T)$ must be close to $f(\mathbf{r})$. Noting that f is bounded, this implies that $E(f(\mathbf{x}_T))$, which equals $\phi(\mathbf{u})$, must also be arbitrarily close to $f(\mathbf{r})$ when **u** approaches **r**.

To exploit this idea, we consider a special case that is suggestive, even though the hypothesis is overly restrictive.

Proposition 2. *If S is* convex, *then $\phi(\mathbf{u}) \to f(\mathbf{r})$ as $\mathbf{u} \to \mathbf{r}$ for every boundary point* **r** *of S.*

To prove this proposition, we will show that (4) holds. Let **r** be any boundary point and N any neighborhood of **r**, and suppose that $\mathcal{P}$ is a *supporting hyperplane* to S at **r**.[26] We can choose new coordinates such that **r** becomes the origin and $\mathcal{P}$ is the plane $u_1 = 0$. Because the Brownian transition function (1) is isotropic, the components of $\{x_t\}$ along and perpendicular to the u_1 axis will be independent Brownian motions in one and $k - 1$ dimensions, respectively.

By Theorem 23.2, the component along the u_1 axis (i.e., perpendicular to $\mathcal{P}$) almost surely has both positive and negative fluctuations in arbitrarily small intervals of time. This implies that if **u** is very close to **r**, then the Brownian motion will reach the hyperplane $\mathcal{P}$ (corresponding to $u_1 = 0$) in an extremely short time, with probability approaching 1. Because $\mathcal{P}$ supports S, in the course of reaching $\mathcal{P}$ the motion must first hit ∂S. But since this hit occurs in a very short time, it will take place at a point in the neighborhood N with probability close to 1. Thus (4) does hold, so that $\phi(\mathbf{u}) \to f(\mathbf{r})$ as $\mathbf{u} \to \mathbf{r}$ at any boundary point **r** where S has a supporting hyperplane. For convex sets this means every boundary point, and so the function $\phi(\mathbf{u})$ solves the Dririchlet problem in the strict sense stated above.

The assumption that S is convex is much stronger than necessary for the existence of a solution. It is not hard to show, for example, that (4) holds at every point where the boundary is smooth. Even some rather ugly behavior can be tolerated. For example, if **r** is the vertex of a cone pointing into S,

---

[26] This means a hyperplane containing **r** that leaves S entirely to one side.

in any number of dimensions, (4) must hold as long as the cone encloses a positive solid angle.

But the Brownian motion interpretation of the solution also gives insight into the possibility that *no* strict solution exists. For example, let S consist of the open unit sphere with one radial line—let us say the nonnegative points on the u_1 axis—deleted. If the dimension of the sphere is at least 3, then it can be shown that a Brownian motion started anywhere within the sphere, even at a point on that missing radius, makes its first hit on the boundary ∂S at some point of the unit sphere proper, not on the "whisker" (which, of course, is also part of ∂S). Thus the values assigned to the boundary function f on the whisker are irrelevant to the definition of $\phi(\mathbf{u})$, and there is no need for $\phi(\mathbf{u})$ to be close to $f(\mathbf{r})$ even if $\mathbf{u} = \mathbf{r}$, when $\mathbf{r}$ is a point of the whisker lying inside the unit sphere. The reason for this behavior is that a Brownian motion in two or more dimensions almost surely never returns exactly to its starting point; similarly, it almost surely does not pass through any other given point in its state space.[27] Therefore in three or more dimensions, a Brownian motion will not hit a particular line such as the u_1 axis at any positive time.

Finally, we remark that the deleted radial line can be fattened into a sort of spine pointing into S, in such a way that (4) still does not hold at the tip of the spine. For this to be true, the spine must come to a point exponentially fast. This example is known as the "Lebesgue thorn," and it shows that even when the boundary of S is homeomorphic to a sphere it may contain *irregular points* where (4) fails, and the generalized Dirichlet solution need not converge to the value of the boundary function f at the point.

For a second and last example, we will sketch the relationship of Brownian motion to eigenvalue problems for the Laplace operator, and describe an elegant method due to Mark Kac by which theorems of H. Weyl and T. Carleman can be obtained using this connection.[28] Suppose now that S is a region in two dimensions whose boundary ∂S is a smooth Jordan curve. The classical problem is to study the nonzero functions ϕ that are continuous in $S \cup \partial S$ and satisfy the system

$$\frac{1}{2}\Delta\phi + \lambda\phi = 0 \quad \text{if} \quad \mathbf{u} \in S; \qquad \phi(\mathbf{u}) = 0 \quad \text{if} \quad \mathbf{u} \in \partial S. \qquad (5)$$

---

[27] For a proof, see Lévy.
[28] M. Kac, "Can one hear the shape of a drum?", *Amer. Math. Monthly* 73 (1966), pp. 1–23.

(Again, Δ denotes the Laplace operator.) It is known that there is a sequence of values $\{\lambda_n\}$ of the parameter λ, the *eigenvalues*, for which nonzero solutions ϕ_n of (5) (the *eigenfunctions*) exist, and that the normalized eigenfunctions form an orthonormal basis for $L_2(S)$. These eigenfunctions and eigenvalues can be calculated explicitly for a few special cases of the region S, but to study them in general is an important and difficult problem.[29]

The eigenfunctions and eigenvalues can be used to construct solutions for the related problem of *heat flow* in the region S. This is an initial-value problem; one version asks us to find the function $\psi(t, \mathbf{u})$ that satisfies the conditions

$$\frac{\partial \psi}{\partial t} = \frac{1}{2}\Delta \psi \qquad \text{for } t > 0, \quad \mathbf{u} \in S; \qquad (6)$$

$$\lim_{\mathbf{u} \to \mathbf{r}} \psi(t, \mathbf{u}) = 0 \qquad \text{for } t > 0, \quad \mathbf{r} \in \partial S, \qquad (7)$$

and

$$\lim_{t \to 0+} \psi(t, \mathbf{u}) = h(\mathbf{u}) \qquad \text{for } \mathbf{u} \in S. \qquad (8)$$

The arbitrary (let us assume piecewise continuous) function h in (8) represents the initial temperature distribution. If we expand the function ψ in a Fourier series using the basis $\{\phi_n\}$, it is not hard to see, at least formally, that the solution is

$$\psi(t, \mathbf{u}) = \int_S h(\mathbf{x}) g(t, \mathbf{x}, \mathbf{u}) \, dx_1 \, dx_2, \qquad (9)$$

where the kernel g, called the *fundamental solution*, is given by

$$g(t, \mathbf{x}, \mathbf{u}) = \sum_{n=1}^{\infty} e^{-\lambda_n t} \phi_n(\mathbf{x}) \phi_n(\mathbf{u}). \qquad (10)$$

In the special cases when S is a finite interval in R^1 or a rectangle in R^2, this construction reduces to the standard "separation of variables" solution of the heat problem that uses Fourier sine series to represent the initial condition h.

There is a similar problem where heat flow takes place in the entire space $S = R^k$; then the boundary condition (7) is replaced by the assumption that $\psi(t, \mathbf{u})$ is bounded. (The initial temperature h is also assumed to be

---

[29] For a detailed nonprobabilistic discussion, see Courant and Hilbert (1953), Chapter 6.

bounded.) In this case, we have the explicit solution

$$\psi(t, \mathbf{u}) = \int_{R^k} h(\mathbf{x}) f(t, \mathbf{x}, \mathbf{u}) \, dx_1 \cdots dx_k, \qquad (11)$$

where the kernel f is defined by

$$f(t, \mathbf{x}, \mathbf{u}) = \frac{1}{(2\pi t)^{k/2}} e^{-|\mathbf{x}-\mathbf{u}|^2/2t}. \qquad (12)$$

This kernel is again the fundamental solution of the problem, relating to the general solution (11) in the same way that (10) relates to (9). But now some sort of relation to probability is apparent, since *the fundamental solution f of the initial value problem is the density of the transition probability function* (1) *for Brownian motion* in R^k. It is natural to look for a corresponding probabilistic interpretation of (10).

We don't have to look very far. Suppose that $\{\mathbf{y}_t\}$ is a Brownian motion with initial state $\mathbf{x} \in S$, but with the modification that on reaching the boundary ∂S the process is "killed" (terminates) or "absorbed." In other words, let $\mathbf{y}_t = \mathbf{x}_t$ for $t < T$, where T is the exit time from S defined in (2). For $t \geq T$, the process is left undefined, or it may be assigned to a new, abstract state (not a point of R^k). Then it can be shown that *this modified Brownian motion is a Markov process*, and that just as in the case of unrestricted Brownian motion, *its transition probability function $q_t(\mathbf{x}, E)$ has a density that is none other than the fundamental solution $g(t, \mathbf{x}, \mathbf{u})$ of the heat flow problem* (6), (7), and (8). This relationship between Brownian motion and initial/boundary value problems for the heat equation (6) has been recognized for many years. However, a general understanding of this and related phenomena depends on semigroup theory and other relatively modern developments concerning Markov processes, and is beyond the scope of this book.

Using the connections just described, however, we can obtain Weyl's theorem in a rigorous manner. The key is an idea that Kac called "the principle of not feeling the boundary." This means that for very small values of t, the presence or absence of the "absorbing" boundary ∂S some distance away has a negligible effect on transition probabilities near the point of origin $\mathbf{x}$ of the process. In particular, we assert that

$$g(t, \mathbf{x}, \mathbf{x}) \sim f(t, \mathbf{x}, \mathbf{x}) \quad \text{as} \quad t \to 0, \quad \mathbf{x} \in S. \qquad (13)$$

This claim can be justified using probabilistically plausible comparisons. First of all, because some paths are eliminated and none are added, the introduction of an absorbing boundary can only reduce the probability of a transition; thus

$$q_t(\mathbf{x}, \mathbf{E}) \leq p_t(\mathbf{x}, \mathbf{E}) \tag{14}$$

for all $\mathbf{x} \in S$ and $\mathbf{E} \subset S$. It follows that $q_t(\mathbf{x}, \mathbf{E})$ is absolutely continuous, and that its density $g(t, \mathbf{x}, \mathbf{u})$ is everywhere less than the density f of p_t. But for the same reasons, if a new absorbing barrier is introduced that lies inside ∂S, the corresponding new transition probabilities will be even smaller than $q_t(\mathbf{x}, \mathbf{E})$. We can, for example, construct a square that lies within S and contains the initial point $\mathbf{x}$, make this square an absorbing barrier, and calculate the new transition probabilities explicitly by means of Fourier sine series. The result is a lower bound to complement the upper bound in (14). Using these two bounds, we can readily establish the asymptotic relation (13).

The rest is analysis, not probability. Combining (13) with (12) and (10), we have

$$\sum_{n=1}^{\infty} e^{-\lambda_n t} \phi_n^2(\mathbf{x}) \sim \frac{1}{2\pi t} \quad \text{as } t \to 0. \tag{15}$$

If we integrate over S, this becomes

$$\sum_{n=1}^{\infty} e^{-\lambda_n t} \sim \frac{A}{2\pi t}, \tag{16}$$

where A is the area of S. But the sum in this relation can be written as a Laplace transform, namely,

$$\sum_{n=1}^{\infty} e^{-\lambda_n t} = \int_0^{\infty} e^{-ts} \, dF(s),$$

where $F(s)$ is the number of eigenvalues λ_n which are less than s. We now appeal to Karamata's Tauberian theorem,[30] which relates the asymptotic behavior of the Laplace transform at 0 to the growth of F at ∞. Using (16) the result is

$$F(s) \sim \frac{A}{2\pi} s \quad \text{as } s \to \infty. \tag{17}$$

---

[30] See Widder (1946), Chapter V, Section 4.

In view of the definition of F, (17) is equivalent to

$$\lambda_n \sim \frac{2\pi}{A} n \quad \text{as } n \to \infty. \tag{18}$$

This is Weyl's theorem.

A later result due to Carleman is also nearly in hand. We return to (15) and treat it as we previously did (16), but without the integration over S. The analog of (17) is Carleman's theorem, namely,

$$\sum_{\lambda_n < s} \phi_n^2(\mathbf{x}) \sim \frac{s}{2\pi} \quad \text{as } s \to \infty. \tag{19}$$

Remark. The title of Kac's 1966 paper was a question: "Can one hear the shape of a drum?" What one "hears" are the overtones as the drum vibrates, that is, the sequence of eigenvalues of the boundary value problem (5). The question thus becomes whether or not knowledge of the $\{\lambda_n\}$ determines the form of the region S. As a step in this direction, Weyl's theorem (18), or even (16), shows that at least the *area* of S can indeed be obtained if the eigenvalues are known. Moreover, assuming that S is convex, Kac shows that the length of the circumference ∂S and whether or not S is circular can also be "heard" (i.e., determined from the eigenvalues).

The full question was open for decades, but has recently been settled. In 1992 C. Gordon, D. Webb, and S. Wolpert constructed two different polygonal regions such that the system (5) has exactly the same eigenvalues for each region.[31] The proof, alas, has nothing to do with Brownian motion.

Final Remark. The real point of this section has not been to provide simple proofs for hard theorems, since by the time all the gaps are filled, it is doubtful if any work has been saved in comparison with "classical" methods. Certainly, some new intuition and insights have been obtained. But the main moral is simply that there are deep connections between probability theory, especially Markov processes, and large areas of classical analysis. These connections have led to profound enrichment of both fields.

---

[31] "You cannot hear the shape of a drum," *Bull. Amer. Math. Soc.* 27 (1992), pp. 134–138.

APPENDIX A

Essentials of Measure Theory

Modern probability is based on the theory of measure and integration, and some familiarity with measures, measurable functions, and integrals is essential for its study. However, only a few specific results beyond the basics will be needed for the purposes of this book. These essentials are stated here. The proofs of these theorems, full details of the construction of measures and integrals, and much more can be found in Kolmogorov and Fomin (1975) or in Royden (1988), as well as in the early chapters of Billingsley (1995) and Loeve (1977).

Measure Spaces

A *finite measure space* is the same thing as the *probability space* defined in Section 1, except that the condition $P(\Omega) = 1$ is replaced by $P(\Omega) < \infty$. (We will not need to discuss infinite measures.) A *measurable space* is a set Ω and a σ-field $\mathcal{B}$ of subsets, without any measure function being given. The real line R^1 together with the σ-field of all Borel sets[1] is the leading example of a measurable space.

Given a measurable space $(\Omega, \mathcal{B})$, it is often necessary to construct a measure function P on $\mathcal{B}$ that has some desired properties; a valuable tool for this task is the extension theorem stated in Section 1. One proof of this theorem begins by constructing an *outer measure* defined for *all* subsets of Ω, and proceeds by showing that this outer measure is actually a measure when restricted to the σ-field $\mathcal{B}$ generated by $\mathcal{F}$ (i.e., the smallest σ-field containing $\mathcal{F}$). For example, in constructing Lebesgue measure on the unit

---

[1] The "Borel sets" of R^1 (or of any metric space) are those sets belonging to the smallest σ-field that contains all the open sets.

interval [0, 1], the outer measure is defined as

$$\mu^*(E) = \inf \sum_{k=1}^{\infty} (b_k - a_k), \tag{1}$$

where the infimum is taken over all countable collections of open intervals (a_k, b_k) in [0, 1] whose union contains the set E. The set function μ^* extends the concept of length of an interval, and can be shown to be countably additive on the σ-field of all Borel sets. In the general case of Theorem 1.1, an outer measure can be defined using coverings by countable unions of sets from the field $\mathcal{F}$ in place of the covering by open intervals in (1).

Measurable Functions and Integration

Suppose $(\Omega, \mathcal{B})$ and $(\Omega', \mathcal{B}')$ are any measurable spaces and f is a mapping from Ω into Ω'. Then f is called *measurable* provided that all inverse images of measurable sets are measurable; that is, provided

$$f^{-1}(C) \in \mathcal{B} \quad \text{whenever} \quad C \in \mathcal{B}'. \tag{2}$$

It is easy to see that the composition of two measurable mappings, when defined, is itself measurable.

Remark. The above definition of the measurability of f depends on both of the σ-fields $\mathcal{B}$ and $\mathcal{B}'$. This point requires care, especially when Ω and Ω' are the same set but $\mathcal{B}$ and $\mathcal{B}'$ are distinct. Books on real variable theory, for example, often say that a function from R^1 into R^1 is "Lebesgue-measurable" if the inverse image of every Borel set is a Lebesgue-measurable set (but not necessarily a Borel set). Such a function is a measurable mapping according to our definition from $(R^1, \mathcal{L})$ into $(R^1, \mathcal{B}')$, where $\mathcal{B}'$ is the σ-field of Borel sets and $\mathcal{L}$ the (larger) σ-field of Lebesgue-measurable sets. Note that two such functions cannot be composed as mappings of measurable spaces, and so the above affirmation of the measurability of composite mappings does not apply to the composition of two Lebesgue-measurable functions. However, if f is Lebesgue-measurable while g is actually *Borel*-measurable (i.e., if g^{-1} carries Borel sets into Borel sets), then the function $g(f)$ is indeed Lebesgue-measurable too.

The real-valued, Borel-measurable functions [i.e., measurable mappings defined on $(\Omega, \mathcal{B})$ such that $(\Omega', \mathcal{B}')$ is R^1 with the Borel field] are especially important, since if $(\Omega, \mathcal{B}, P)$ is a probability space, these functions become the "random variables." We will call these functions "Borel-measurable," or, if there is no ambiguity, simply "measurable." To prove measurability in this situation it is not necessary to check condition (2) for every Borel set in R^1; it suffices to verify it for all open intervals of the form (a, ∞). An important "meta-theorem" about these functions states roughly that any new function obtained from a countable set of them by algebraic and/or limiting operations is again measurable. In particular we note the following.

Proposition 1. *If $f_1, f_2, \ldots$ are Borel-measurable, so are $\sup f_n$, $\inf f_n$, $\limsup f_n$, and $\liminf f_n$, as well as $\lim f_n$ and $\sum f_n$ if they exist.*

To define the general Lebesgue integral, suppose that $(\Omega, \mathcal{B}, P)$ is a finite measure space and that f is a real-valued function on Ω. If f takes only the values 0 and 1, it is the *indicator function* of the set $A = \{\omega \in \Omega : f(\omega) = 1\}$ (denoted $\mathbf{1}_A$), and f will be measurable if and only if A is a measurable set (i.e., if $A \in \mathcal{B}$). In this case, we define $\int f\, dP = P(A)$. The definition extends to simple functions (finite linear combinations of indicator functions) in the obvious way:

$$\text{if } \quad f(\omega) = \sum_{k=1}^{n} c_k \mathbf{1}_{A_k}(\omega) \quad \text{then} \quad \int f\, dP = \sum_{k=1}^{n} c_k P(A_k).$$

Now suppose that f is a nonnegative measurable function and define

$$\int_\Omega f\, dP = \sup\left\{\int g\, dP : g \text{ simple}, g \leq f\right\}. \tag{3}$$

If the supremum in (3) is finite, the function f is said to be *integrable*. Finally, a general real-valued function can be decomposed into the difference of two nonnegative functions:

$$f = f^+ - f^- = \max(f, 0) + \min(f, 0).$$

Then f is integrable provided both f^+ and f^- are integrable, and we set

$$\int_\Omega f\, dP = \int_\Omega f^+\, dP - \int_\Omega f^-\, dP.$$

This completes the definition. Of course, it remains to be shown that the integral so defined extends the definition for simple functions given earlier, and satisfies such basic properties as linearity and nonnegativity (see Section 2).

Convergence Theorems

It is frequently important to know whether the operations of integration and passage to a limit can be interchanged. The Lebesgue integral just defined has desirable properties in this respect.

Theorem 1 (Monotone Convergence). *Suppose that f_n, $n = 1, 2, \ldots$, is a sequence of nonnegative measurable functions on the measure space $(\Omega, \mathcal{B}, P)$ such that $f_n \leq f_{n+1}$ for all n. Then*

$$\lim_{n \to \infty} \int_\Omega f_n \, dP = \int_\Omega \lim_{n \to \infty} f_n \, dP \leq \infty. \tag{4}$$

Theorem 2 (Dominated Convergence). *Suppose that f_n are measurable functions on $(\Omega, \mathcal{B}, P)$ and that $\lim f_n(\omega) = f(\omega)$ exists for all ω. Suppose there exists an integrable function g on the same space such that $|f_n(\omega)| \leq g(\omega)$ for all ω. Then again*

$$\lim_{n \to \infty} \int_\Omega f_n \, dP = \int_\Omega f \, dP. \tag{5}$$

Theorem 3 (Fatou's Lemma). *Suppose f_n are nonnegative measurable functions on $(\Omega, \mathcal{B}, P)$ and that again $\lim f_n(\omega) = f(\omega)$ exists. Then f is integrable provided $\liminf \int_\Omega f_n \, dP < \infty$, and in this case we have*

$$\int_\Omega f \, dP \leq \liminf_{n \to \infty} \int_\Omega f_n \, dP. \tag{6}$$

The first two of these results will be used frequently in this book. Fatou's lemma is needed only once; see Problem 17.5.

Product Spaces and Fubini's Theorem

Suppose we are given two finite measure spaces $(\Omega, \mathcal{B}, P)$ and $(\Omega', \mathcal{B}', P')$. Let $\Omega \times \Omega'$ denote the direct product of Ω and Ω', that is, the set of all ordered pairs (ω, ω'), where $\omega \in \Omega$ and $\omega' \in \Omega'$. A *rectangle* is a set in $\Omega \times \Omega'$ that is itself the direct product of some set $E \in \mathcal{B}$ and some $E' \in \mathcal{B}'$. The smallest σ-field that contains all rectangles is called the *product σ-field* and denoted by $\mathcal{B} \times \mathcal{B}'$.

There is a natural way to define a measure Π, called the *product measure*, on the measurable space $(\Omega \times \Omega', \mathcal{B} \times \mathcal{B}')$. For rectangles, the product measure Π is simply

$$\Pi(E \times E') = P(E) P'(E'). \tag{7}$$

The set of all finite unions of disjoint rectangles is a field, and Π can be defined for such sets by adding the measures (7) of the rectangles. The extension theorem (Theorem 1.1) then guarantees the existence of a measure function Π defined on $\mathcal{B} \times \mathcal{B}'$ that extends (7); this is the *product measure*. The measure space $(\Omega \times \Omega', \mathcal{B} \times \mathcal{B}', \Pi)$ is called the *product* of $(\Omega, \mathcal{B}, P)$ and $(\Omega', \mathcal{B}', P')$. The most familiar example, of course, is the relationship of Lebesgue measure in the plane R^2 to the linear Lebesgue measures on the x and y axes.

The principal theorem about these measures relates the integral of a function f over the direct product to the integrals of related functions over the individual (factor) spaces; this is the result that says that "double integrals" and "iterated integrals" are equal.

Theorem 4 (Fubini). *Suppose $f = f(\omega, \omega')$ is a measurable, real-valued function on the product space $(\Omega \times \Omega', \mathcal{B} \times \mathcal{B}')$, and that f is integrable with respect to the product measure Π. Then*

$$\int_{\Omega \times \Omega'} f(\omega, \omega') \, d\Pi = \int_{\Omega} \left(\int_{\Omega'} f(\omega, \omega') \, dP' \right) dP$$

$$= \int_{\Omega'} \left(\int_{\Omega} f(\omega, \omega') \, dP \right) dP'.$$

If either of the two iterated integrals exists with f replaced by $|f|$, then f is integrable with respect to Π and the equality holds.

It is implicit in this theorem that the "inner integrals" (the terms in parentheses) exist for almost all values of the variable.

The construction of a product space can easily be extended to more than two factors. If the number of factors is finite, the construction given above need hardly be changed. If the number is infinite, suppose the factors are $(\Omega_n, \mathcal{B}_n, P_n)$ where $n = 1, 2, \ldots$; for simplicity, we assume these are probability spaces [i.e., that $P_n(\Omega_n) = 1$ for all n]. We now use rectangles of the form

$$R = (E_1 \times E_2 \times \cdots \times E_k \times \Omega_{k+1} \times \Omega_{k+2} \times \cdots);$$

that is, the product set is restricted on a finite number of coordinates. These rectangles are called *cylinder sets*. The product measure is defined in the natural way for such sets as

$$\Pi(R) = \prod_{j=1}^{k} P_j(E_j).$$

Again finite unions of cylinder sets make up the field $\mathcal{F}$, and the extension theorem applies. The construction carries through even when the number of factors is not countable, but we will not need to consider this case.

Product measures and Fubini's theorem are frequently useful in connection with independent random variables as seen in Section 3 and elsewhere. However, the theorem also has some less immediate applications (i.e., Section 23).

Absolute Continuity and the Radon–Nikodym Theorem

Suppose that $(\Omega, \mathcal{B})$ is a measurable space, and that P and Q are both finite measures on this space. The measure Q is said to be *absolutely continuous* with respect to P, often denoted $Q \ll P$, if $P(E) = 0$ implies $Q(E) = 0$. It is easy to see that this must happen if Q is defined by

$$Q(E) = \int_E f \, dP, \tag{8}$$

where f is a nonnegative function integrable with respect to P. The converse is also true.

Theorem 5 (Radon–Nikodym). *Suppose that P and Q are finite measures on $(\Omega, \mathcal{B})$ and that Q is absolutely continuous with respect to P. Then there exists a nonnegative, P-integrable function f such that (8) holds for all $E \in \mathcal{B}$.*

The function f is called the *Radon–Nikodym derivative* or simply the *density* of the measure Q. This theorem will be needed only for proving the existence of the conditional expectation of a random variable (Section 4).

Bibliography

P. Billingsley, *Convergence of Probability Measures* (New York: Wiley, 1968).
P. Billingsley, *Probability and Measure,* 3rd ed. (New York: Wiley, 1995).
R. Courant and D. Hilbert, *Methods of Mathematical Physics* (New York: Wiley-Interscience, 1953–1962).
W. Feller, *An Introduction to the Theory of Probability and Its Applications,* Vol. 1, 3rd ed. (New York: Wiley, 1968).
W. Feller, *An Introduction to the Theory of Probability and Its Applications,* Vol. 2 (New York: Wiley, 1966).
J. Galambos, *The Asymptotic Theory of Extreme Order Statistics,* 2nd ed. (Malabar, FL: Krieger, 1987).
B. V. Gnedenko and A. N. Kolmogorov, *Limit Distributions for Sums of Independent Random Variables,* 2nd ed. (Reading, MA: Addison-Wesley, 1968), translated by K. L. Chung.
A. N. Kolmogorov, *Foundations of the Theory of Probability,* 2nd ed. (New York: Chelsea, 1956).
A. N. Kolmogorov and S. V. Fomin, *Introductory Real Analysis* (New York: Dover, 1975), translated by Richard Silverman.
J. Lamperti, *Stochastic Processes: A Survey of the Mathematical Theory* (New York: Springer-Verlag, 1977).
P. Lévy, *Processus Stochastiques et Mouvement Brownien,* 2nd ed. (Paris: Gauthier-Villars, 1965).
M. Loeve, *Probability Theory,* 4th ed. (New York: Springer-Verlag, 1977).
S. Port and C. Stone, *Brownian Motion and Classical Potential Theory* (New York: Academic Press, 1978).
H. Royden, *Real Analysis,* 3rd ed. (New York: Macmillan, 1988).
W. Rudin, *Real and Complex Analysis,* 3rd ed. (New York: McGraw-Hill, 1987).
F. Spitzer, *Principles of Random Walk* (Princeton: Van Nostrand, 1964).
D. V. Widder, *The Laplace Transform* (Princeton: Princeton Univ. Press, 1946).

Index

Absolutely continuous measures, 2, 9, 180 (see also Radon-Nikodym theorem)
Absorbing barrier (for Brownian motion), 171
Almost-sure convergence (see Convergence of random variables)
Attraction (see Domain)
Avogadro's number (and Brownian motion), 132

Bachelier, L., 131, 158
Bernoulli, James, 31
Bernoulli trials, 31, 35
    and characteristic functions, 82, 83
Bernstein polynomials, 38
Berry, A. C., 99
Billingsley, P., 162, 163, 175
Binomial distribution, 31
Borel, E., 41, 43
Borel–Cantelli lemma, 44
Borel fields, 1
    generated by a family of sets, 3
    generated by a family of random variables, 20, 57
    generated by cylinder sets, 29, 156
    generated by rectangles, 27, 154, 178
    independent, 58–59
Borel sets (of a metric space), 71, 161, 175
Brown, Robert, 131
Brownian motion (phenomenon), 131
Brownian motion path functions, 143–149
    law of iterated logarithm for, 143
    local law of iterated logarithm, 146
    nondifferentiability, 135, 145–146, 147
    nonrectifiability, 149
    zeros of, 145, 147
Brownian motion process, 131–149
    absorbing barrier for, 171
    construction of, first, 132–141
        second, 156
        third, 163
    definition of, 132
    derivative of, 135–136
    eigenvalue problems and, 169–173

Brownian motion process (*Continued*)
   heat flow and, 171
   potential theory and, 165–169
   several dimensions, 166
   transition function of, 153, 166

Carleman, T., 169, 173
Cauchy distribution, 87–88, 101
   domain of attraction of, 113
Central limit theorem, 82, 94–100
   for densities, 100
   speed of convergence, 99
   unequal distributions, 99
   unit circle (random variables on), 97
Central limit problem, 81–82
Chandrasekhar, S., 101
Chapman–Kolmogorov equation, 153
Characteristic functions, 82–90
   convergence of measures and (see Continuity theorem)
   definition, 82–83
   density function and, 87
   higher dimensions, 92–94
   inversion formula for, 84–87
   moments and (see Moments)
   symmetric distributions and, 86
Characteristic sequence, 96
Chebyshev, P. L., 32, 82
Chebyshev's inequality, 33
Chung, K. L. and Fuchs, W. H. J., 126, 130
Ciesielski, Z., 133
Complete probability measure, 4–5
Compound Poisson distributions, 121
   and infinitely divisible distributions, 122

Conditional expectation, 18, 20–25
   definition of, 20–21
   existence of, 21–22
   properties, 23
Conditional independence, 18
Conditional probability, 17, 21, 22
Conditionally compact (family of measures), 77–78
Conditioning, 19
Consistency (for finite-dimensional distributions), 28, 155
Continuity condition for measures, 2
Continuity theorem,
   for characteristic functions, 84, 88, 89
   for characteristic sequences (on the unit circle), 97
   for Laplace transforms, 91
Continuous case (of a random variable), 9
Continuous function,
   approximation by polynomials (see Weierstrass)
Convergence in distribution (of stochastic processes), 158
Convergence of measures (see Weak convergence)
Convergence of random series (see Random series)
Convergence of random variables (definitions),
   almost sure, 10
   in the mean, 11
   in probability, 11
Convergence theorems for Lebesgue integral, 178
Convergence hull of a random sample, 117–118
Convex sets and the Dirichlet problem, 168

INDEX

Convolution, 16–17
   characteristic functions and, 83
   continuity with respect to weak convergence, 110
Correlation coefficient, 35–36
Courant, R. and Hilbert, D., 166, 170
Covariance, 35–36
Covariance function (of Brownian motion), 133, 140
Covariance matrix, 37, 141
Cramer, H., 99
Cylinder sets, 29, 179 (see also Rectangles)

De Moivre limit theorem, 82, 83, 100, 158
Delta function, 135
Density (see Probability density)
Diagonal method, 76
Diffusion process, 157
Dirichlet problem, 166–169
   generalized solution to, 167
   irregular boundary points, 169
Distribution function, 3–4, 7
   normal (see Normal distribution)
   maximum of random variables, 78
   sum of random variables, 16
   weak convergence and, 72–74
Distribution (measure),
   absolutely continuous, 9
   of a random variable, 7
   of several random variables (see Joint distributions)
   of a sum of independent random variables, 16–17
Domain of attraction (see Stable)
Domain of maximal attraction (see Maximal)

Dominated convergence theorem, 178
Donsker, M. D., 161
Doob, J. L., 11
Dwass, M., 165

Eigenvalue problems (for the Laplace operator), 169–173
Einstein, Albert, 131, 132
Erdös, Paul M., 161, 163
Esseen, C. C., 99
Expected value, 5–9
   conditional (see Conditional expectation)
   independent random variables and, 12–14, 60
   properties of, 6
   ways of computing, 6–9, 15
Exponential distribution, 119
Extension theorem for measures, 2–3
Extremal process, 165

Fatou's lemma, 178
Feller, W., vii, 79, 99, 126, 150, 163
Finite-dimensional distributions (of a stochastic process), 28, 29–30
   consistency conditions for, 27–28, 155
Fisher, L., 118
Fisher, R. A. and Tippett, L. H. C., 81, 114
Fomin, S. V., 175
Frechet, M., 81
Friedrichs, K. O., 166
Fubini's theorem, 179
   applications of, 16, 148

Fuchs, W. H. J. (see Chung, K. L.)
Fundamental solution, 170, 171

Galambos, J., 81, 118
Generalized functions, 135
Generating function, 108, 127, 128
Gnedenko, B. V., 81, 114, 117
Gnedenko and Kolmogorov, 38, 57, 99, 100, 113, 121
Gordon, C., Webb, D., and Wolpert, S., 173
Groups (sums of random variables on), 96, 98

Haar functions, 136–137
  completeness, 137–138
Haar measure, 98
Hardy, G. H. and Littlewood, J. E., 62, 65
Harmonic function, 166, 167
Hausdorff, F., 62
Heat equation, 157, 170
Helly–Bray theorem, 72
Helly's theorem, 74, 76
Hilbert, D. (see Courant)
Hilbert space (L_2 space), 22, 48
Holtzmark, J., 101

Independent Borel fields (see Borel fields)
Independent events, 12
Independent increments, 132
Independent random variables, 12–17
Index (see Stable distributions)
Indicator function, 5, 177

Infinitely divisible distributions, 119–125
  characteristic function of, 122–125
  definition, 119
  limit theorems and, 118–121
  weak convergence and, 122
Interval recurrence, 126
Invariance principle, 161
Inversion formula (see Characteristic functions)
Irregular point (see Dirichlet problem)
Iterated logarithm (law of), 62, 67–70
  for Brownian motion (see Brownian motion path functions)

Joint characteristic function (see Characteristic functions)
Joint distribution (of several random variables), 15 (see also Finite-dimensional distributions)

Kac, Mark, 161, 163, 169, 171, 173
Kakutani, S., 167
Kampe de Feriet, J., 133
Karamata, J., 79, 172
Khintchine, A., 62, 67, 69, 124
Kolmogorov, A. N., 3, 12, 175
  convergence of random series, 53
  existence of random variables, 27–29, 154
  infinitely divisible laws, 124
  law of iterated logarithm, 62, 70

INDEX   **187**

Kolmogorov's inequality, 46, 60
zero–one (0–1) law, 57, 60

L distributions, 118
L_2 space (see Hilbert space)
Lamperti, J., 147, 156, 157, 160, 164–165, 166
Laplace transform, 91–92
  convergence of measures and (see Continuity theorem)
Laplace operator, 166, 169–170
Law of iterated logarithm (see Iterated logarithm)
Law of large numbers (see Strong law, Weak law)
Lebesgue integral, 5, 175–176, 177 (see also Expected value)
Lebesgue thorn, 169
Lévy, Paul, 101
  Brownian motion, 143, 146, 169
  infinitely divisible laws, 124
  limiting distributions and stable laws, 112
Levy distance (between distributions), 74
  weak convergence and, 75
Lewy, H., 166
Lindeberg, J. W., 99
Lindeberg condition for central limit theorem, 99, 100
Littlewood, J. E. (see Hardy)
Local limit theorems, 100
Loeve, M., 57, 175
Lyapunov, A. M., 82

Markov chains, 126, 150, 153–154
Markov process, 150–157
  definition, 150

Markov property, 108, 150–152
Markov transition functions, 152–154, 157
  Brownian motion, 153, 166, 171
Maximal distributions, 114–117
  definition, 114
  domain of maximal attraction, 117
  limit theorems and, 114
Maximum partial sum,
  inequalities for, 46, 53, 66–67
  limiting distribution for, 158–160
Maximum of a random sample, 78
  limiting distributions for, 79–81, 114
  normally distributed, 81
Mean (see Expected value)
Measurability, 5, 15, 176, 177
  of a stochastic process, 148
Measurable space, 7, 175
Measure space, 175
Moments (of a random variable), 32
  characteristic functions and, 94, 107
Monotone convergence theorem, 178

Nisio, M., 136
Normal distribution, 10, 17, 69, 87
  central limit theorem, 82, 89, 95, 99–100
  characteristic function of, 83, 142
  joint (several dimensions), 25, 141–143
  maximum of a sample, 81

Normal distribution (*Continued*)
 moments, 10
 self-reproducing property, 101
Normal numbers, 43

Ornstein, L. S., 146
Outer measure, 175–176

Parseval's relation, 137
Perfect probability spaces, 11
Persistence (see Recurrence)
Poisson distribution, 118, 119
 compound, 121–122
Poisson limit theorem, 92
Poisson process, 153
Polya, George, 126
Port, S. and Stone, C., 166
Potential theory (and Brownian motion), 166–169
Principle of not feeling the boundary, 171
Probability density, 9
 convolution and, 17
 obtained from characteristic function, 87
Probability measure, 1–4
 complete, 4–5
 extension theorem for, 3
Probability space, definition, 1
 construction, 2–4, 26–29
 perfect, 11
Product measure, 26–27, 178–179 (see also Fubini's theorem)
Projection operator, 22
Prokhorov, Yu. V., 78, 161–162, 163
Pugwash Conferences, vii
Pythagorean theorem, 33

Rademacher functions, 62
Radon–Nikodym theorem, 180
 applications of, 21
Random series, 47, 52–57, 134, 136, 138
 power series, 61
Random sums, 19, 33, 121
Random variable (definition), 5, 177
 convergence (see Convergence of random variables)
 existence of, 26–29
Random walk, 108, 126, 128, 157–160
Rectangles (measure of), 26
Recurrence (for sums of independent random variables), 125–130
Recurrence time (for random walk), 108
Reflection principle (for simple random walk), 159–160
Regular variation, 79
Responsibility of scientists, viii
Rotblat, Joseph, viii
Royden, H., viii, 175
Rudin, W., 166, 167

Sampling, 36–37
Schauder functions, 137, 138
Self-reproducing property (see Stable)
Self-similar (stochastic process), 164–165
Semi-stable (stochastic process), 164–165
Sigma field or σ-field (see Borel field)
Slowly varying functions, 79, 113

Spitzer, Frank, 130
Stable distributions, 101–109
  definition, 103
  domain of attraction, 113
  index, 104, 107
  limit theorems and, 110, 112
  self-reproducing property, 102, 103
  symmetric, 104
Standard deviation, 32
Stochastic matrix, 153, 154
Stochastic process, 27–28
Stone, C. (see Port)
Strong law of large numbers, 41–43, 48, 50–52
Strong Markov property, 147, 160, 167
Symmetrization (of a distribution), 107, 111

Tail field (of a sequence of random variables), 60, 61
Three series theorem, 52–53
Tight family of measures, 78
Tippett, L. H. C. (see Fisher)
Transience, 126
Triangular array (of random variables), 118
  limit theorems for, 118–120
Truncated normal distribution, 159–160
Truncation method, 50, 52–53
Type (of a distribution), 9, 103
  weak convergence and, 111

Uhlenbeck, G. E., 146
Uncorrelated random variables, 35
Uniqueness theorem
  for characteristic functions, 84–86, 93
  for characteristic sequences, 96
  for Laplace transforms, 91
  for weak limits, 73
Unit circle (random variables on), 96–98

Variance, 32, 33

Weak convergence (of measures), 71–78
  definition, 71
Weak law of large numbers, 31, 34–36, 38
Webb, D. (see Gordon)
Weierstrass approximation theorem, 38–41
Weyl, Hermann, 169, 173
Widder, D. V., 172
Wiener, Norbert, 132, 134, 147
  Dirichlet problem and, 166
Wiener measure, 162
Wiener process (see Brownian motion)
Wolpert, S. (see Gordon)

Zero–one (0–1) law (see Kolmogorov)

WILEY SERIES IN PROBABILITY AND STATISTICS

ESTABLISHED BY WALTER A. SHEWHART AND SAMUEL S. WILKS

Editors
Vic Barnett, Ralph A. Bradley, Nicholas I. Fisher, J. Stuart Hunter, J. B. Kadane, David G. Kendall, David W. Scott, Adrian F. M. Smith, Jozef L. Teugels, Geoffrey S. Watson

Probability and Statistics
  ANDERSON · An Introduction to Multivariate Statistical Analysis, *Second Edition*
  *ANDERSON · The Statistical Analysis of Time Series
  ARNOLD, BALAKRISHNAN, and NAGARAJA · A First Course in Order Statistics
  BACCELLI, COHEN, OLSDER, and QUADRAT · Synchronization and Linearity:
    An Algebra for Discrete Event Systems
  BARTOSZYNSKI and NIEWIADOMSKA-BUGAJ · Probability and Statistical Inference
  BERNARDO and SMITH · Bayesian Statistical Concepts and Theory
  BHATTACHARYYA and JOHNSON · Statistical Concepts and Methods
  BILLINGSLEY · Convergence of Probability Measures
  BILLINGSLEY · Probability and Measure, *Second Edition*
  BOROVKOV · Asymptotic Methods in Queuing Theory
  BRANDT, FRANKEN, and LISEK · Stationary Stochastic Models
  CAINES · Linear Stochastic Systems
  CAIROLI and DALANG · Sequential Stochastic Optimization
  CHEN · Recursive Estimation and Control for Stochastic Systems
  CONSTANTINE · Combinatorial Theory and Statistical Design
  COOK and WEISBERG · An Introduction to Regression Graphics
  COVER and THOMAS · Elements of Information Theory
  CSÖRGÖ and HORVÁTH · Weighted Approximations in Probability Statistics
  *DOOB · Stochastic Processes
  DUDEWICZ and MISHRA · Modern Mathematical Statistics
  ETHIER and KURTZ · Markov Processes: Characterization and Convergence
  FELLER · An Introduction to Probability Theory and Its Applications, Volume 1,
    Third Edition, Revised; Volume II, *Second Edition*
  FREEMAN and SMITH · Aspects of Uncertainty: A Tribute to D. V. Lindley
  FULLER · Introduction to Statistical Time Series, *Second Edition*
  FULLER · Measurement Error Models
  GIFI · Nonlinear Multivariate Analysis
  GUTTORP · Statistical Inference for Branching Processes
  HALD · A History of Probability and Statistics and Their Applications before 1750
  HALL · Introduction to the Theory of Coverage Processes
  HANNAN and DEISTLER · The Statistical Theory of Linear Systems
  HEDAYAT and SINHA · Design and Inference in Finite Population Sampling
  HOEL · Introduction to Mathematical Statistics, *Fifth Edition*
  HUBER · Robust Statistics
  IMAN and CONOVER · A Modern Approach to Statistics
  JUREK and MASON · Operator-Limit Distributions in Probability Theory
  KAUFMAN and ROUSSEEUW · Finding Groups in Data: An Introduction to Cluster
    Analysis
  LAMPERTI · Probability: A Survey of the Mathematical Theory, *Second Edition*
  LARSON · Introduction to Probability Theory and Statistical Inference, *Third Edition*
  LESSLER and KALSBEEK · Nonsampling Error in Surveys
  LINDVALL · Lectures on the Coupling Method
  MANTON, WOODBURY, and TOLLEY · Statistical Applications Using Fuzzy Sets

*Now available in a lower priced paperback edition in the Wiley Classics Library.

Probability and Statistics (Continued)
- MARDIA · The Art of Statistical Science: A Tribute to G. S. Watson
- MORGENTHALER and TUKEY · Configural Polysampling: A Route to Practical Robustness
- MUIRHEAD · Aspects of Multivariate Statistical Theory
- OLIVER and SMITH · Influence Diagrams, Belief Nets and Decision Analysis
- *PARZEN · Modern Probability Theory and Its Applications
- PRESS · Bayesian Statistics: Principles, Models, and Applications
- PUKELSHEIM · Optimal Experimental Design
- PURI and SEN · Nonparametric Methods in General Linear Models
- PURI, VILAPLANA, and WERTZ · New Perspectives in Theoretical and Applied Statistics
- RAO · Asymptotic Theory of Statistical Inference
- RAO · Linear Statistical Inference and Its Applications, *Second Edition*
- *RAO and SHANBHAG · Choquet-Deny Type Functional Equations with Applications to Stochastic Models
- RENCHER · Methods of Multivariate Analysis
- ROBERTSON, WRIGHT, and DYKSTRA · Order Restricted Statistical Inference
- ROGERS and WILLIAMS · Diffusions, Markov Processes, and Martingales, Volume I: Foundations, *Second Edition;* Volume II: Îto Calculus
- ROHATGI · An Introduction to Probability Theory and Mathematical Statistics
- ROSS · Stochastic Processes
- RUBINSTEIN · Simulation and the Monte Carlo Method
- RUBINSTEIN and SHAPIRO · Discrete Event Systems: Sensitivity Analysis and Stochastic Optimization by the Score Function Method
- RUZSA and SZEKELY · Algebraic Probability Theory
- SCHEFFE · The Analysis of Variance
- SEBER · Linear Regression Analysis
- SEBER · Multivariate Observations
- SEBER and WILD · Nonlinear Regression
- SERFLING · Approximation Theorems of Mathematical Statistics
- SHORACK and WELLNER · Empirical Processes with Applications to Statistics
- SMALL and McLEISH · Hilbert Space Methods in Probability and Statistical Inference
- STAPLETON · Linear Statistical Models
- STAUDTE and SHEATHER · Robust Estimation and Testing
- STOYANOV · Counterexamples in Probability
- STYAN · The Collected Papers of T. W. Anderson: 1943–1985
- TANAKA · Time Series Analysis: Nonstationary and Noninvertible Distribution Theory
- THOMPSON and SEBER · Adaptive Sampling
- WELSH · Aspects of Statistical Inference
- WHITTAKER · Graphical Models in Applied Multivariate Statistics
- YANG · The Construction Theory of Denumerable Markov Processes

Applied Probability and Statistics
- ABRAHAM and LEDOLTER · Statistical Methods for Forecasting
- AGRESTI · Analysis of Ordinal Categorical Data
- AGRESTI · Categorical Data Analysis
- AGRESTI · An Introduction to Categorical Data Analysis
- ANDERSON and LOYNES · The Teaching of Practical Statistics
- ANDERSON, AUQUIER, HAUCK, OAKES, VANDAELE, and WEISBERG · Statistical Methods for Comparative Studies
- ARMITAGE and DAVID (editors) · Advances in Biometry
- *ARTHANARI and DODGE · Mathematical Programming in Statistics
- ASMUSSEN · Applied Probability and Queues

*Now available in a lower priced paperback edition in the Wiley Classics Library.

Applied Probability and Statistics (Continued)
  *BAILEY · The Elements of Stochastic Processes with Applications to the Natural Sciences
  BARNETT and LEWIS · Outliers in Statistical Data, *Second Edition*
  BARTHOLOMEW, FORBES, and McLEAN · Statistical Techniques for Manpower Planning, *Second Edition*
  BATES and WATTS · Nonlinear Regression Analysis and Its Applications
  BECHHOFER, SANTNER, and GOLDSMAN · Design and Analysis of Experiments for Statistical Selection, Screening, and Multiple Comparisons
  BELSLEY · Conditioning Diagnostics: Collinearity and Weak Data in Regression
  BELSLEY, KUH, and WELSCH · Regression Diagnostics: Identifying Influential Data and Sources of Collinearity
  BERRY · Bayesian Analysis in Statistics and Econometrics: Essays in Honor of Arnold Zellner
  BERRY, CHALONER, and GEWEKE · Bayesian Analysis in Statistics and Econometrics: Essays in Honor of Arnold Zellner
  BHAT · Elements of Applied Stochastic Processes, *Second Edition*
  BHATTACHARYA and WAYMIRE · Stochastic Processes with Applications
  BIEMER, GROVES, LYBERG, MATHIOWETZ, and SUDMAN · Measurement Errors in Surveys
  BIRKES and DODGE · Alternative Methods of Regression
  BLOOMFIELD · Fourier Analysis of Time Series: An Introduction
  BOLLEN · Structural Equations with Latent Variables
  BOULEAU · Numerical Methods for Stochastic Processes
  BOX · R. A. Fisher, the Life of a Scientist
  BOX and DRAPER · Empirical Model-Building and Response Surfaces
  BOX and DRAPER · Evolutionary Operation: A Statistical Method for Process Improvement
  BOX, HUNTER, and HUNTER · Statistics for Experimenters: An Introduction to Design, Data Analysis, and Model Building
  BROWN and HOLLANDER · Statistics: A Biomedical Introduction
  BUCKLEW · Large Deviation Techniques in Decision, Simulation, and Estimation
  BUNKE and BUNKE · Nonlinear Regression, Functional Relations and Robust Methods: Statistical Methods of Model Building
  CHATTERJEE and HADI · Sensitivity Analysis in Linear Regression
  CHATTERJEE and PRICE · Regression Analysis by Example, *Second Edition*
  CLARKE and DISNEY · Probability and Random Processes: A First Course with Applications, *Second Edition*
  COCHRAN · Sampling Techniques, *Third Edition*
  *COCHRAN and COX · Experimental Designs, *Second Edition*
  CONOVER · Practical Nonparametric Statistics, *Second Edition*
  CONOVER and IMAN · Introduction to Modern Business Statistics
  CORNELL · Experiments with Mixtures, Designs, Models, and the Analysis of Mixture Data, *Second Edition*
  COX · A Handbook of Introductory Statistical Methods
  *COX · Planning of Experiments
  COX, BINDER, CHINNAPPA, CHRISTIANSON, COLLEDGE, and KOTT · Business Survey Methods
  CRESSIE · Statistics for Spatial Data, *Revised Edition*
  DANIEL · Applications of Statistics to Industrial Experimentation
  DANIEL · Biostatistics: A Foundation for Analysis in the Health Sciences, *Sixth Edition*
  DAVID · Order Statistics, *Second Edition*
  *DEGROOT, FIENBERG, and KADANE · Statistics and the Law
  *DEMING · Sample Design in Business Research
  DILLON and GOLDSTEIN · Multivariate Analysis: Methods and Applications

  *Now available in a lower priced paperback edition in the Wiley Classics Library.

Applied Probability and Statistics (Continued)

DODGE and ROMIG · Sampling Inspection Tables, *Second Edition*

DOWDY and WEARDEN · Statistics for Research, *Second Edition*

DRAPER and SMITH · Applied Regression Analysis, *Second Edition*

DUNN · Basic Statistics: A Primer for the Biomedical Sciences, *Second Edition*

DUNN and CLARK · Applied Statistics: Analysis of Variance and Regression, *Second Edition*

ELANDT-JOHNSON and JOHNSON · Survival Models and Data Analysis

EVANS, PEACOCK, and HASTINGS · Statistical Distributions, *Second Edition*

FISHER and VAN BELLE · Biostatistics: A Methodology for the Health Sciences

FLEISS · The Design and Analysis of Clinical Experiments

FLEISS · Statistical Methods for Rates and Proportions, *Second Edition*

FLEMING and HARRINGTON · Counting Processes and Survival Analysis

FLURY · Common Principal Components and Related Multivariate Models

GALLANT · Nonlinear Statistical Models

GLASSERMAN and YAO · Monotone Structure in Discrete-Event Systems

GREENWOOD and NIKULIN · A Guide to Chi-Squared Testing

GROSS and HARRIS · Fundamentals of Queueing Theory, *Second Edition*

GROVES · Survey Errors and Survey Costs

GROVES, BIEMER, LYBERG, MASSEY, NICHOLLS, and WAKSBERG · Telephone Survey Methodology

HAHN and MEEKER · Statistical Intervals: A Guide for Practitioners

HAND · Discrimination and Classification

*HANSEN, HURWITZ, and MADOW · Sample Survey Methods and Theory, Volume 1: Methods and Applications

*HANSEN, HURWITZ, and MADOW · Sample Survey Methods and Theory, Volume II: Theory

HEIBERGER · Computation for the Analysis of Designed Experiments

HELLER · MACSYMA for Statisticians

HINKELMAN and KEMPTHORNE: · Design and Analysis of Experiments, Volume 1: Introduction to Experimental Design

HOAGLIN, MOSTELLER, and TUKEY · Exploratory Approach to Analysis of Variance

HOAGLIN, MOSTELLER, and TUKEY · Exploring Data Tables, Trends and Shapes

HOAGLIN, MOSTELLER, and TUKEY · Understanding Robust and Exploratory Data Analysis

HOCHBERG and TAMHANE · Multiple Comparison Procedures

HOCKING · Methods and Applications of Linear Models: Regression and the Analysis of Variables

HOEL · Elementary Statistics, *Fifth Edition*

HOGG and KLUGMAN · Loss Distributions

HOLLANDER and WOLFE · Nonparametric Statistical Methods

HOSMER and LEMESHOW · Applied Logistic Regression

HØYLAND and RAUSAND · System Reliability Theory: Models and Statistical Methods

HUBERTY · Applied Discriminant Analysis

IMAN and CONOVER · Modern Business Statistics

JACKSON · A User's Guide to Principle Components

JOHN · Statistical Methods in Engineering and Quality Assurance

JOHNSON · Multivariate Statistical Simulation

JOHNSON and KOTZ · Distributions in Statistics

               Continuous Univariate Distributions—2

               Continuous Multivariate Distributions

JOHNSON, KOTZ, and BALAKRISHNAN · Continuous Univariate Distributions, Volume 1, *Second Edition*

JOHNSON, KOTZ, and BALAKRISHNAN · Discrete Multivariate Distributions

*Now available in a lower priced paperback edition in the Wiley Classics Library.

Applied Probability and Statistics (Continued)
  JOHNSON, KOTZ, and KEMP · Univariate Discrete Distributions, *Second Edition*
  JUDGE, GRIFFITHS, HILL, LÜTKEPOHL, and LEE · The Theory and Practice of
      Econometrics, *Second Edition*
  JUDGE, HILL, GRIFFITHS, LÜTKEPOHL, and LEE · Introduction to the Theory and
      Practice of Econometrics, *Second Edition*
  JUREČKOVÁ and SEN · Robust Statistical Procedures: Asymptotics and Interrelations
  KADANE · Bayesian Methods and Ethics in a Clinical Trial Design
  KADANE AND SCHUM · A Probabilistic Analysis of the Sacco and Vanzetti Evidence
  KALBFLEISCH and PRENTICE · The Statistical Analysis of Failure Time Data
  KASPRZYK, DUNCAN, KALTON, and SINGH · Panel Surveys
  KISH · Statistical Design for Research
  *KISH · Survey Sampling
  LAD · Operational Subjective Statistical Methods: A Mathematical, Philosophical, and
      Historical Introduction
  LANGE, RYAN, BILLARD, BRILLINGER, CONQUEST, and GREENHOUSE ·
      Case Studies in Biometry
  LAWLESS · Statistical Models and Methods for Lifetime Data
  LEBART, MORINEAU., and WARWICK · Multivariate Descriptive Statistical
      Analysis: Correspondence Analysis and Related Techniques for Large Matrices
  LEE · Statistical Methods for Survival Data Analysis, *Second Edition*
  LePAGE and BILLARD · Exploring the Limits of Bootstrap
  LEVY and LEMESHOW · Sampling of Populations: Methods and Applications
  LINHART and ZUCCHINI · Model Selection
  LITTLE and RUBIN · Statistical Analysis with Missing Data
  MAGNUS and NEUDECKER · Matrix Differential Calculus with Applications in
      Statistics and Econometrics
  MAINDONALD · Statistical Computation
  MALLOWS · Design, Data, and Analysis by Some Friends of Cuthbert Daniel
  MANN, SCHAFER, and SINGPURWALLA · Methods for Statistical Analysis of
      Reliability and Life Data
  MASON, GUNST, and HESS · Statistical Design and Analysis of Experiments with
      Applications to Engineering and Science
  McLACHLAN and KRISHNAN · The EM Algorithm and Extensions
  McLACHLAN · Discriminant Analysis and Statistical Pattern Recognition
  McNEIL · Epidemiological Research Methods
  MILLER · Survival Analysis
  MONTGOMERY and MYERS · Response Surface Methodology: Process and Product
      in Optimization Using Designed Experiments
  MONTGOMERY and PECK · Introduction to Linear Regression Analysis, *Second Edition*
  NELSON · Accelerated Testing, Statistical Models, Test Plans, and Data Analyses
  NELSON · Applied Life Data Analysis
  OCHI · Applied Probability and Stochastic Processes in Engineering and Physical
      Sciences
  OKABE, BOOTS, and SUGIHARA · Spatial Tesselations: Concepts and Applications
      of Voronoi Diagrams
  OSBORNE · Finite Algorithms in Optimization and Data Analysis
  PANKRATZ · Forecasting with Dynamic Regression Models
  PANKRATZ · Forecasting with Univariate Box-Jenkins Models: Concepts and Cases
  PORT · Theoretical Probability for Applications
  PUTERMAN · Markov Decision Processes: Discrete Stochastic Dynamic Programming
  RACHEV · Probability Metrics and the Stability of Stochastic Models
  RÉNYI · A Diary on Information Theory
  RIPLEY · Spatial Statistics

  *Now available in a lower priced paperback edition in the Wiley Classics Library.

Applied Probability and Statistics (Continued)
  RIPLEY · Stochastic Simulation
  ROSS · Introduction to Probability and Statistics for Engineers and Scientists
  ROUSSEEUW and LEROY · Robust Regression and Outlier Detection
  RUBIN · Multiple Imputation for Nonresponse in Surveys
  RYAN · Modern Regression Methods
  RYAN · Statistical Methods for Quality Improvement
  SCHUSS - Theory and Applications of Stochastic Differential Equations
  SCOTT · Multivariate Density Estimation: Theory, Practice, and Visualization
  SEARLE · Linear Models
  SEARLE · Linear Models for Unbalanced Data
  SEARLE · Matrix Algebra Useful for Statistics
  SEARLE, CASELLA, and McCULLOCH · Variance Components
  SKINNER, HOLT, and SMITH · Analysis of Complex Surveys
  STOYAN, KENDALL, and MECKE · Stochastic Geometry and Its Applications, *Second Edition*
  STOYAN and STOYAN · Fractals, Random Shapes and Point Fields: Methods of Geometrical Statistics
  THOMPSON · Empirical Model Building
  THOMPSON · Sampling
  TIERNEY · LISP-STAT: An Object-Oriented Environment for Statistical Computing and Dynamic Graphics
  TIJMS · Stochastic Modeling and Analysis: A Computational Approach
  TITTERINGTON, SMITH, and MAKOV · Statistical Analysis of Finite Mixture Distributions
  UPTON and FINGLETON · Spatial Data Analysis by Example, Volume 1: Point Pattern and Quantitative Data
  UPTON and FINGLETON · Spatial Data Analysis by Example, Volume II: Categorical and Directional Data
  VAN RIJCKEVORSEL and DE LEEUW · Component and Correspondence Analysis
  WEISBERG · Applied Linear Regression, *Second Edition*
  WESTFALL and YOUNG · Resampling-Based Multiple Testing: Examples and Methods for *p*-Value Adjustment
  WHITTLE · Optimization Over Time: Dynamic Programming and Stochastic Control, Volume I and Volume II
  WHITTLE · Systems in Stochastic Equilibrium
  WONNACOTT and WONNACOTT · Econometrics, *Second Edition*
  WONNACOTT and WONNACOTT · Introductory Statistics, *Fifth Edition*
  WONNACOTT and WONNACOTT · Introductory Statistics for Business and Economics, *Fourth Edition*
  WOODING · Planning Pharmaceutical Clinical Trials: Basic Statistical Principles
  WOOLSON · Statistical Methods for the Analysis of Biomedical Data
  *ZELLNER · An Introduction to Bayesian Inference in Econometrics
Tracts on Probability and Statistics
  BILLINGSLEY · Convergence of Probability Measures
  KELLY · Reversability and Stochastic Networks
  TOUTENBURG · Prior Information in Linear Models